T0290620

CONSTRUCTION DISPUTES

CONSTRUCTION DISPUTES
Seeking Sensible Solutions

Wayne Clark

LONDON PUBLISHING PARTNERSHIP

Contents

Contents

A Tribute to Stephen Hepburn

My book is dedicated to Stephen Hepburn, who I met in Interlaken, Switzerland, in July 2020. At this, our first – and, as it turned out, only – meeting, we enjoyed lunch and a few glasses of wine. During the rather long lunch, Stephen encouraged me to write a book on construction contracts and claims. He encouraged me to write *this* book.

We kept in touch regularly after our Interlaken lunch, chatting on the phone, exchanging emails and messages, strategizing over my developing book.

Sadly, Stephen passed away suddenly on 8 January 2021. I miss him very much.

Rest in peace my friend.

Foreword by Husam Gawish

There are very few active professionals in the construction industry who are as experienced or as well equipped in the area of construction claims and dispute resolution as Wayne Clark. Readers of this unique book have the opportunity to benefit from Wayne's considerable experience in the field – experience that has primarily focused on dispute avoidance and finding *sensible solutions*, a theme that runs right through his excellent book. Wayne has managed to skillfully draw on his fifty years of practical, hands-on involvement with claims management and dispute resolution to deliver this considerable and important contribution to the industry he loves and has been a part of for so long. This simple and easy-to-follow guide is a truly valuable resource to help you find a 'sensible solution' to construction disputes.

Disputes in construction are nothing new, and nor will they ever go away. They are inevitable. They occur for a multitude of reasons that are not always obvious and cannot be foreseen. Disputes differ in nature depending on contracts, region, culture and legal jurisdiction. Common causes of construction disputes include changes in scope, contract interpretation issues, contract administration failures, incomplete or incorrect design information and site access restrictions. By no means are the causes limited to these examples, either: the list is long and extensive.

Disputes also arise when projects are awarded on the basis of price and not quality. Contractors find themselves engaged in pricing wars. Not only is the budget underestimated but often the execution programme is unachievable. The project is doomed to failure before it even begins. The can is simply kicked down the road, paving the way for inevitable future claims and disputes. Other factors also come into play, such as the effectiveness and independence of the engineer, the willingness of the employer to be reasonable, the experience and quality of the contractor and, more importantly, the environment in which the contracting parties are operating. Is everyone singing from the same hymn sheet? Is there a common-sense approach of togetherness and a genuine will by the parties to cooperate and collaborate in order to achieve a common goal?

Construction contracts – in all their numerous forms and types, whether standard or bespoke – are intended to administer and manage the execution of a project. The contract aims to ensure that the project works are delivered in accordance with the agreed scope, within budget and within the agreed timeframe. This sounds simple in theory, but in reality it is seldom a straightforward process. No matter how well written a contract is, it simply cannot cater for every situation or eventuality. More often than not, projects encounter events and issues that ultimately lead to delay, disruption and additional costs. The impact on the successful delivery of the project will vary depending on the nature of the events. Disputes naturally arise when the parties disagree and argue about where the blame and responsibility for those events lie. Construction contracts require the parties to give timely notices of events and claims. Contracts also provide a procedure for the evaluation and assessment of these claims and for awarding additional time and/or costs if entitlement can be demonstrated. If it is not possible for the parties to reach agreement, the contract will also provide dispute-resolution mechanisms that allow the parties to engage in formal proceedings, whether these are in the form of litigation, arbitration or any other alternative dispute-resolution procedure (e.g. mediation, expert determination or dispute boards).

Disputes can run for extended periods of time, reaching several years in many cases. More importantly, they cost a significant amount of money, and the damage inflicted on relationships is often irreparable. This is precisely why Wayne's book is an invaluable guide – one that offers *sensible solutions* to the construction industry. The book provides a clear step-by-step guide to all parties on

- best practice in managing projects;
- how to avoid disputes in the first place;

- drafting strong, credible and persuasive claims;
- the art and skill of negotiation; and
- the ideal process for managing the resolution of disputes.

Having had the privilege of working closely with Wayne for more than four years and having known him for longer, I have first-hand knowledge of his extraordinary ability to distil the facts and focus on the crux of the salient issues. Wayne's fifty years of experience in the construction industry, working for both contractors and consultants in different parts of the world, is brought to the fore in this must-read book. Wayne's focus on avoiding unnecessary conflict, and on finding that sensible solution, is central to the advice and guidance he gives throughout his book. As he notes in his preface, he tries 'to focus on ways to avoid disputes, or to manage the disputes when they cannot be avoided'. Wayne has certainly achieved that goal.

The book also looks at the dilemmas faced by parties as a result of the Covid-19 pandemic, and Wayne includes recommendations for achieving a fair and sensible outcome. He also looks briefly at third-party funding and how this can encourage the parties to settle. An unexpected inclusion in the book is the introduction of the 'shadow arbitrator', who, if commissioned early in the dispute process, can guide parties and legal counsel to prepare claims and arbitral pleadings that will persuade a tribunal – and encourage the parties to find that sensible solution.

Wayne's book will be of particular interest to young construction professionals who are in the early stages of their careers. It will help them avoid the pitfalls of poor project management. It will inspire them to strengthen relationships and avoid disputes, and it will guide them on how best to manage disputes if all else fails. More seasoned construction practitioners will no doubt find Wayne's book a valuable reminder and reference that they can fall back on when the wheels of their projects start becoming loose.

Thankfully, Wayne remains very much active in construction and will no doubt have much more to offer to construction professionals of all ages and from all corners of the globe in the years to come.

Preface

Disputes are not uncommon in the construction world. In my experience, construction disputes more often than not arise because of the human factor – because of people. Initial differences of opinion become disputes, and these can develop into serious conflict. Once an impasse is reached – when it seems impossible for the parties to agree on anything – the disputing parties often believe the only path left for final resolution is arbitration or litigation.

Fortunately, most people are sensible and usually find solutions to their differences. Solutions that avoid time-consuming and costly arbitration or litigation. Solutions that can prevent irreparable damage to relationships.

Disputes can be avoided at an early stage, when issues and problems first surface – issues that are affecting, or may go on to affect, the efficient performance of a project. How these situations are managed will determine whether or not the differences can be overcome and serious conflict avoided.

In this book I try to focus on ways to avoid disputes, or how to manage the disputes when they cannot be avoided. I deal with the art of developing credible construction claims – claims that will encourage, even persuade, the parties to engage in constructive dialogue and find that sensible solution. My book also briefly looks at the process of resolving disputes when negotiations fail, at resolving disputes through third-party resolution such as mediation, and dispute adjudication boards. And for those occasions when third-party resolution fails, I discuss the final stage for resolution: arbitration, which is the final option available in most construction contracts. Very few of the projects I have been involved in have ended up in the courts. The vast majority are referred to arbitration for final resolution. It is for this reason that the focus in my book is on arbitration for final resolution of construction disputes.

While I explore various ways to manage and resolve disputes in the book, a large part of it deals with the management and preparation of credible claims – claims that will persuade an arbitral tribunal. If the claim is good enough to persuade a tribunal, a wise recipient should recognize this and be persuaded to seek and find a sensible solution.

Later in the book I discuss the role of the shadow arbitrator. Shadow arbitrators can help parties and legal counsel prepare claims and arbitration pleadings that will persuade a tribunal and that should also persuade the other party to settle. I also briefly look into third-party funding and how this might also encourage parties to settle.

With regard to preparing claims, I have assumed – in fact, I highly recommend – that the preparation of contractors' claims is undertaken by an independent consultant who is an expert in dispute and claims management – someone with hands-on experience in settlement negotiations and who understands the dispute-resolution process. Ideally, one wants to find someone who has experience in arbitration or is a practising arbitrator – someone who can think like an arbitrator.

The reason I encourage contractors to engage an independent consultant/expert is for their objectivity. While contractors' in-house personnel might be perfectly capable of preparing claims, personalities, emotions and ownership of the issues all get in the way. Weaknesses may not be viewed objectively – or worse, they might be overlooked or ignored.

In-house personnel often focus on what they *perceive* to be undisputed entitlement – their *perceived* rights – whether or not they have the evidence to support such perceptions. Weaknesses that are overlooked or ignored are weaknesses that will certainly be exploited by the other side. Unless they are recognized and addressed, such weaknesses will turn out to be the other side's strengths.

In line with my views on why in-house personnel should not draft their own claims, this book is written on the premise that contractors' claims will be undertaken by a team of independent construction consultants, all experienced and/or experts in their own particular fields. When developing the claims, the team will work closely with the contractor, their client, to agree on the structure and content of the claim and to agree a strategy for reaching an early and amicable settlement. That is, they will look for a strategy for success.

Throughout the book I refer to the contractor (who could equally be a subcontractor) as the 'contractor', 'the client', 'our client' or 'my client'. The independent consultant is me, and my support team is referred to as 'our claims team', 'my team' or 'our team'. Where reference is made to 'the other party' or 'the other side', this relates to the employer (the building owner) and to the employer's representatives (the engineer, architect or project manager). These representatives are collectively referred to under the title 'employer'.

Concerning 'my team', in the forty-or-so years I have been involved in construction dispute management, I have worked with many people, from varying backgrounds. Some were experts in their respective disciplines and most of them were highly competent. While living in Doha, Qatar, I had the pleasure – indeed the privilege – of working with four exceptional ladies. We worked together as a team on several commissions, preparing successful claims for happy clients. I was proud to be part of this extraordinary team. Thank you Shirley Mendoza, Emelyn Martinez, Lynsey Bromley and Nurul Sabri.

Acknowledgements

My sincere thanks and gratitude go to the following people and organizations who kindly gave me permission to refer to and quote from their books or articles, and for guidance on matters outside of my expertise.

Julian Bailey. Quotes from Julian's *Construction Law* (3rd edition), which was published by London Publishing Partnership in 2020.

Professor Dr Jörg Risse. Quotes from Jörg's article 'The shadow arbitrator: mere luxury or real need?', which was published in *ASA Bulletin*, volume 38, issue 2 (June 2020).

Tony Bingham. Quotes from Tony's article 'Blow your rights', which was published in *Building* magazine on 9 May 2003.

Paulina Touroude. Paulina, who is the head of the enforcement risk division and a senior member of the legal analysis team at Profile Investment in France, provided guidance on the third-party funding section of the book.

Corrs Chambers Westgarth. This Sydney, Australia law firm kindly gave me permission to make reference to their article 'Time for Australia to embrace dispute resolution boards', written by Lucy Goldsmith and Andrew Stephenson.

The Society of Construction Law (SCL). Quotes from the SCL's *Delay and Disruption Protocol* (2nd edition), which was published by the society in 2017.

The International Federation of Consulting Engineers (FIDIC).
Express permission was granted by FIDIC's chief executive officer
to quote from FIDIC Conditions of Contract for Construction
(1999 Red Book) and Conditions of Contract for Plant and
Design-Build (1999 Yellow Book), the copyright of which is
acknowledged. Permission was also granted to quote from their
April 2020 publication 'FIDIC guidance note: COVID-19 guidance memorandum to users of FIDIC standard forms of works
contract'.

I was also given permission to quote from the *2018 Rules on the
Efficient Conduct of Proceedings in International Arbitration (Prague
Rules)*, which are available online.

⁓

I would also like to thank my friends and colleagues Shirley Mendoza, Emelyn Martinez, Nurul Sabri, Lynsey Bromley, Christopher
Seppälä, Rob Valenta and David Archer for allowing me to mention
them by name in various parts of my book.

⁓

Finally, to my late friend Stephen Hepburn: without his initial idea
and constant encouragement, I would never have written this book.
Thank you Stephen.

CONSTRUCTION DISPUTES

CHAPTER 1

Managing Construction Projects to Prevent Disputes

*All parties to a construction project will have the same objective:
complete the project to a high standard, on time and within budget.
Efficient performance will minimize disputes and ensure good relationships are maintained.*

Managing construction projects to prevent disputes is about getting the job done and getting it done well.

Careful planning and good performance by all parties are pre-
requisites for the success of any construction project. By all parties I
mean the employer, employer's representatives, contractors, subcontractors and suppliers. Planning and performance are essential for all
phases of a project, from the employer's concept (or dream) through
to completion and handing over the works.

The various construction project phases we will discuss in this
chapter, each with a focus on dispute avoidance, are

(i) employer's concept and planning,
(ii) the tender (bidding) phase,
(iii) the construction phase and
(iv) the employer's taking over of the completed works.

Two other important areas of construction management that
are necessary to help prevent or minimize disputes are wise contract
management and managing people. We will look at managing people
later in this chapter and discuss wise contract management in the next.

EMPLOYER'S CONCEPT AND PLANNING

When an employer wishes to engage in a construction project – to build something – be it a private development or the expansion of a country's infrastructure, the employer will usually appoint an architect or engineer to convert the employer's idea, its concept, into detailed design drawings and project documents. These drawings and project documents will form the basis of the contractor's scope, price and time for executing and completing the project works.

Thorough planning by the employer and meticulous preparation of the project documents by the employer's team (architect, engineer, quantity surveyor, etc.) are fundamental to the success of any construction project. It is essential for the employer to allow its team enough time to prepare the necessary documents, ensuring that they are well thought through and, more importantly, that they are complete.

An important decision to be made by the employer early in the initial planning stage is the type of contract to be used for its project. A clear understanding by the employer and its team of the various forms of contract available, and the importance of choosing the right contract, can be a deciding factor in whether a project runs smoothly or runs into trouble.

A few years ago, I was invited to conduct a workshop for an employer's board of directors on the various forms of contract it was considering for its new project. The board members could not agree on the allocation of risk to be borne by the contractor. Some members advocated that the contractor should bear all the risks, while others were concerned how this might impact on their limited budget. The contracts they were debating over were three of the FIDIC 1999 suite of contracts (books): namely, the Red Book, the Yellow Book and the Silver Book.*

---

*The Red Book: Conditions of Contract for Construction. The Yellow Book: Conditions of Contract for Plant and Design-Build. The Silver Book: Conditions of Contract for EPC/Turnkey Projects.

The workshop lasted three days. All three of the FIDIC contracts, plus a few other standard forms of contract I introduced, were examined in detail. The discussions were constructive and we all learned a considerable amount about the various contracts that would or would not work for the project. We also discussed the importance of the contract documentation being well structured, unambiguous and, most of all, complete.

The contract eventually selected by the board included a balanced allocation of risk. The project documents prepared by the employer's team were clear and explicit regarding scope and risk, and, importantly, the documents were complete before tenders were invited. The result? The project was completed on time, was conflict free and the final cost was within the employer's budget.

Trouble is often born at the planning (concept) stage of a construction project. Rushing to start a project without allowing sufficient time to properly prepare, having insufficient budget, disproportionate risk allocation, or allotting inadequate time for executing the construction works are all ingredients for trouble. They are recipes for failing to achieve the primary objective of all parties: a project that is completed on time, within budget and without conflict.

Another project I worked on some years ago is an example of rushed planning and unrealistic budgeting by the employer. Initial tenders were invited on a design-build basis, with the contractor bearing the vast majority of risks. All bids received by the employer were significantly over budget. There was little time or hope for the employer – the government – to be granted any additional funds for the project. Panic ensued. This was an extremely important and sensitive project for the employer. Emergency meetings were held with the employer's project team (architect, engineer and quantity surveyor). What to do? Hasty decisions were made. The architect was to amend the tender drawings, reducing some areas of the building. This would save money. The engineer and the quantity surveyor would rewrite the contract, changing it from a design-build to a build-only contract, transferring most of the risks to the employer. This should encourage the bidders to lower their prices. It worked.

Retendering took place with the new concept. The lowest bid was within the employer's original budget. Construction began and the employer proudly announced this prestigious facility would be completed and made available to the people by the original opening date. When making this promise, the employer made no allowance for the time it took to redesign the facility, to redraft the project documents or to undertake the retendering process.

The trouble had only just begun.

As the employer's original concept was a design-build project, the drawings were conceptual only, to be developed by the contractor's design team. The employer's decision to change to a build-only project meant the architect had to develop the detailed design and prepare the drawings necessary for the contractor to execute the works. As a result, there were not enough drawings available for the contractor to start building. Later, it was discovered that many of the redesigned areas of the building did not work. The architect had to correct its design and issue new drawings to the contractor.

As expected, the project works were severely delayed and numerous claims followed. The final project cost was almost three times the employer's initial budget. Disputes were plentiful and the project was delivered to the people two years later than promised.

Careful planning by the employer, well-prepared project documents and prudent decisions regarding appropriate risk allocation are essential for the success of all construction projects.

THE TENDER (BIDDING) PHASE

Preparing and compiling tender documents, most of which will become an integral part of the contract, are central to the success of construction projects. Well-drafted documents are fundamental for preventing uncertainty, misunderstanding or misinterpretation – and, more importantly, for preventing disputes. Poorly prepared documents are a real cause of most construction disputes.

The documents provided by the employer at the tender (bidding) stage should, at the very least, include

- conditions of contract (general and particular conditions*);
- specifications;
- drawings;
- schedules, such as bills of quantities (for build-only contracts), schedule of rates, etc.;
- employer's requirements (scope of project works);
- site data and any other available information, such as subsurface conditions and existing underground services.

A complete set of documents will be provided with the employer's 'instructions to tenderers'. These instructions should describe, in intimate detail, the procedure and directives to be followed by the tenderers (bidders) when preparing and submitting their tenders (bids).

As observed in the earlier example of the problems caused by hastily prepared documents, the importance of well-considered, properly prepared and complete project documents cannot be overstated. In my experience, the vast majority of disputes in construction projects have their roots in inadequate project documents.

The first and most important action of the tenderer (the bidder) is to scrutinize each and every document to ensure they

(i) are complete according to the documents listed in the invitation to tender;
(ii) are clear as to the contractor's scope and obligations;

---

*The 'Particular Conditions of Contract' relate to the specific requirements of each project and may amend, replace or delete some of the standard terms and conditions found in the 'General Conditions'. If there is any conflict between the General Conditions and the Particular Conditions, the Particular Conditions will prevail.

(iii) do not contain contradictions, either within or between the documents; and

(iv) contain all information and data necessary to price and execute the works.

If any errors or omissions are noted or if there is ambiguity and/or contradiction in any of the documents, the bidders must immediately notify the employer in writing, requesting any missing information and/or clarification regarding ambiguity or contradiction.

The employer has a duty to promptly respond to requests for clarification, further information or correction. Bidders are advised to be wary of vague responses from the employer. When vague or unhelpful responses are received – something that is not uncommon in my experience – early warning bells should start ringing. There will be trouble ahead.

This brings me to another important exercise during the bidding phase: the risk analysis.

In my early days working for an international contractor I was introduced to the risk analysis. I had recently been employed by this company as a contract manager and was immediately given the task of preparing the risk analysis for a large design-build project in Europe. I had to learn, and learn quickly. Risk analyses were undertaken by this company for each and every tender, and the analyses were extremely thorough.

The key aspects of the project that were examined in our risk analysis included

- project documents,
- employer's requirements,
- specifications,
- design (being a design-build contract),
- contract terms/conditions and the governing law,
- the client and
- the location of the project.

The risks, having been identified, were categorized into levels of importance: high, moderate and low risks. Senior management decided whether or not the risks could be accepted and absorbed, or if the risks were to be valued and added to the tender price. The risks were never ignored.

Getting the price right

Getting the price right is always difficult for the decision makers, particularly if the project is of great importance.

While I was working for a construction company in London, the dream of the owner and managing director was to win the refurbishment of one of London's famous bridges over the river Thames. This bridge was refurbished every seven or so years and was due to be refurbished again that particular year. Our company had received an invitation to tender for the refurbishment works.

The main elements of the works to be priced were labour, materials (mainly paint), stone cleaning and access scaffolding. There were no drawings or bills of quantities to help calculate the quantity of materials required so we had to guess the quantity of paint needed for the entire project. Scaffolding and stone-cleaning prices were received from subcontractors. The most important element of our price – the critical one – was the cost of labour needed to execute the works.

David Archer (the construction company's owner and managing director), Eddie (the operations director) and I deliberated over the price to be included for labour. This was key to winning or losing the project. We agonized over the labour cost for several days. As David was desperate to win the project, he insisted the labour price was to be kept as low as possible. Too low, I argued. We finally agreed on a cost that, in my view, was sensible. We submitted our price. David could not sleep for days, convinced that our price was too high.

The big day arrived. Tenders were opened at 12 noon. We won! David was beside himself with joy. Then came the phone call from the employer. An urgent meeting was called. The employer was concerned. Our price was too low. The fear was that we would not be able to execute and complete the works for the price. If we could not convince the employer's quantity surveyor that our price was right, the contract would be awarded to the next highest bidder, whose price, we learned, was 20% higher than ours.

Fortunately, I was able to persuade the quantity surveyor and we were finally awarded the contract. To the relief of all parties, we completed on time and exactly on budget. We got the price right, but it was a nerve-wrecking experience for all concerned.

Contractors sometimes submit low bids intentionally – even loss-making bids – simply to win the project in the hope of recovering their losses through additional-cost claims. This strategy invariably fails and more often than not leads to unjustified claims and disputes.

When, and only when, the bidders are satisfied that all documents have been carefully considered, that the risks have been provided for and that their price is right should their bid be submitted.

Two-stage tendering (bidding) process

The two-stage bidding process is becoming increasingly popular for international construction projects.

In the first stage of the process the employer develops its design, budget and outline programme. Based on this preliminary information, the bidders submit their method statement, proposed programme, overhead/profit mark-up and a price for indirect costs (site overheads, insurances, performance bonds, etc. – known collectively in the trade as 'preliminaries').

During the second stage of the process the employer and 'preferred bidder' negotiate and agree the contract price, the terms and conditions of contract and the programme.

An advantage of this two-stage process is that the employer and contractor collaborate throughout the second stage of the bidding phase, which should improve the chances of a dispute-free project.

THE CONSTRUCTION PHASE

The contract has been signed and the celebrations are over – it is now time for the contractor to begin preparing for and planning the construction phase of the project. Careful preparation and planning at this early stage is vital for a smooth, trouble-free project.

Assembling the contractor's team

A wide range of skills is needed for the successful execution of construction projects: technical skills (engineers and designers), experienced executive and site managers, supervisors, contract managers, planners, surveyors, administration and support staff. Just as important, if not more important, are the skills needed to carry out the works: good carpenters, bricklayers, electricians, steel fixers and many other trades are required to carry out the actual construction works.

When assembling the project team, the decision makers should ensure suitable staff allocation is made available to maintain and manage the contractor's records. Accurate records are essential for monitoring progress, productivity and costs. Reliable records are also vital for supporting claims. Without the records – without the evidence – claims will most certainly fail. Good record-keeping will be discussed a little later in this chapter.

The key personnel necessary to ensure the contract terms and procedures are complied with are contract administrators (managers). Contract managers are vital to help prevent disputes and to maintain good working relationships with all parties.

When planning the construction team for a project in Greece, the construction manager delegated to lead this particular project questioned the need for, as he put it, 'this expensive contract manager'. The previous project he had managed – which had included this expensive contract manager against his wishes – was completed with no contractual disputes whatsoever. 'So why the need to burden my budget with this superfluous contract person?' he enquired.

'The answer is quite simple,' said the company's managing director. 'There were no disputes on your project simply because you had this expensive contract manager on your team. He was able to diffuse potential disputes using sound negotiating skills and finding sensible solutions.'

This expensive contract manager went on to form part of the construction team for that Greek project, despite the construction manager's initial objections.

A well thought out and balanced construction team is essential for the efficient and effective execution of any construction project.

Obligations and responsibilities under the contract

Understanding the terms and conditions of the contract is fundamental for survival in the construction world. Understanding all aspects of the contract, in particular each party's obligations, is crucial for avoiding conflict and for the smooth, trouble-free execution of the works.

All construction contracts will stipulate the parties' obligations – both general and specific. It surprises me how often contractors overlook, misunderstand or are unaware of their basic contractual obligations. Instead, they focus on their rights. It also surprises me how often employers' representatives (engineers, project managers, etc.) also ignore or are unaware of their obligations and those of the employer. This combination of both the employer and the contractor ignoring or being unaware of their respective obligations has been the cause of many disputes.

Employer's obligations

Typical employer obligations that are sometimes overlooked or not adequately managed include the following.

- Providing the contractor with access to and possession of the project site.
- Ensuring acquisition of the land required for the project.
- Obtaining the necessary permits and licenses for undertaking the project works.
- Ensuring accuracy and completeness of designs, drawings and all other project documents.

- Providing adequate subsurface data.
- Instructing changes/variations in a timely manner.
- Approving the contractor's drawings, programme, materials, suppliers, subcontractors, etc., in a timely manner.
- Fair and timely payment to the contractor for works executed.
- Taking over the works.

Not complying with these basic employer obligations often results in delay, disrupted working, additional cost and disputes.

Cash flow is vital for the survival of any organization, and nowhere is that more true than in the construction industry. A common failure of many employers is not fulfilling their obligation relating to payment. Delayed approval of progress payments, under-certification for works properly executed and finding excuses to refuse taking over completed works to avoid releasing retention moneys are regular features on the many projects I have been involved in.

The failure to fulfill these important obligations has, in some projects I have worked on, been a deliberate tactic. These tactics can, and often do, create severe cash-flow problems for contractors, and more often than not they lead to conflict, and at times bankruptcies.

Contractor's obligations

Contractors have obligations that are equally important for trouble-free projects. Typical obligations that are often overlooked or ignored by contractors include the following.

- Commencement of the works. Most, if not all, contracts stipulate that the contractor has to start work within a specified time from the project commencement date, failing which any resultant delay will be the contractor's responsibility.
- Operations on site. The contractor is to provide adequate resources (labour, plant and equipment) for the proper and timely execution of the project works.
- Progress of the works. To ensure that the project completion date is met and that any intermediate milestones are achieved.

- Design of the project works. Under design-build contracts (e.g. the FIDIC Yellow Book*), the contractor is responsible for the design of the project works.
- Giving timely notice of delay and/or additional-cost claims. Many contracts will deny all contractors' claims for time and/or money if the contractor fails to submit written notice of its claim within the time stipulated in the contract.
- Mitigation of delays. Irrespective of the cause of, or liability for, any delay, the contractor will have an obligation to reduce the effect of the delay on the project works.

A common misconception some contractors have under design-build contracts is believing they will be entitled to additional payment and more time for changes in the design.

A contractor client of mine that was engaged on a major design-build infrastructure project in the Middle East insisted that my team prepare a claim for several design changes that resulted in considerable additional costs. When investigating the design changes, we discovered that the contractor's design team had decided to improve the employer's concept without consulting either the employer or the contractor's management. The design team had assumed the contractor would be paid for any additional costs resulting from their design 'improvements'.

Our client was not happy with my report. There would be no entitlement to additional payment for the design changes. Under the terms of the contract, the contractor was entirely responsible for the final design, including any changes to the employer's concept design.

Furthermore, my report warned that the contractor would be liable for delay-damages should any additional work resulting from the design changes delay completion.

---

*The FIDIC Yellow Book is the 'Conditions of Contract for Plant and Design-Build': for electrical and mechanical plant, and for building and engineering works, designed by the contractor.

My client – wisely, in my opinion – decided against filing the claim, thereby preventing an inevitable dispute with the employer.

Contractor's rights

Apart from its many obligations under the contract, the contractor also has rights. Basic rights such as regular and timely payment for works executed and fair compensation for delays and delay-related costs resulting from events for which the employer is responsible. Many disputes – in fact, the vast majority of disputes I have witnessed – relate to contractors or subcontractors being deprived of these basic rights.

In many cases contractors received payment three to four months after the works in question had been completed, and quite often works properly executed have been deliberately under-certified.

Such undesirable tactics, whether by employers or by contractors when short-changing subcontractors, place a heavy burden on cash flow. These tactics can be detrimental to the efficient performance of the project works and they can increase the likelihood of serious conflict. Such tactics have, in at least one case I am aware of, resulted in termination and, unsurprisingly, arbitration.

On an infrastructure project I worked on in Europe, a serious dispute arose between the contractor and one of its subcontractors. The main issue in dispute was underpayment by the contractor. The contractor was deliberately withholding significant amounts from the subcontractor's monthly progress payments. This created a severe cash-flow problem for the subcontractor, which in turn impacted on the subcontractor's performance.

The subcontractor's complaints were ignored by the contractor's site management, leading to prolonged and heated arguments. Emotions were running high, and they exploded when the subcontractor complained to the contractor's head office. Soon thereafter the contractor's site management terminated the subcontract, citing poor performance by the subcontractor to justify termination.

> The subcontractor filed for arbitration. During the arbitral proceedings the tribunal learned of the contractor's deliberate underpayment and the effect it had on the subcontractor's performance. The tribunal found that termination of the subcontract was unjust and awarded the subcontractor substantial damages.

When the parties' rights and obligations are understood and complied with, the chances of a successful and trouble-free project are greatly enhanced.

The budgets

At the outset of a project the contractor's site and bidding managers should work together to analyse the contract price, establish the net cost of the works (having set aside the overhead and profit mark-up and any risk contingencies), and allocate the net cost to the various elements of the works. The main elements being

- indirect costs (site establishment, site management and support staff, etc.) and
- direct costs (labour, materials, plant and equipment, subcontractors, and suppliers).

Having identified the net costs for the various elements, these costs will form the site management's budgets for the works – budgets that must be closely monitored throughout the project. Identifying budget overruns in good time will give site management the opportunity to take immediate action to rectify the situation. Budget overruns will also alert site management to potential claims.

The risks

As discussed in the 'tender (bidding) phase' section earlier, one of the contractor's tasks is to prepare the risk analysis, identifying the risks and their level of importance.

The risk analysis must be made available to and discussed with the contractor's site management when the contract has been awarded.

Once aware of the risks, the site management will be able to develop a strategy to monitor and effectively manage them.

The programme

The construction manager must, in my opinion, take the lead in planning the project works. It is the construction manager who will understand the complexities of the various stages of the project and they should work with the planners to develop a realistic programme – one that can be achieved with the available resources.

In recent years I have noticed when working with certain international construction companies that the planners, and not the construction managers, dictate the planned durations for the various construction activities. To me, this is the tail wagging the dog. It makes little or no sense for planners – many of whom have little or no experience in procurement or managing construction works – to take the lead in developing the programme. On some projects I have been involved with, where the programme was led by the planners, there were significant delays and severe cost overruns due to inadequate planning and resource allocation of key construction activities.

Resource-loaded programmes are essential tools for planning and organizing the project works, and for monitoring and managing the budgets. The lack of a properly resourced programme – or programmes that are not resourced at all, which is not uncommon in my experience – will create serious problems for contractors when managing the resources and when preparing claims for unproductive or disrupted working.

The project programme should be as uncomplicated as possible. While some construction projects can be complex, the majority of projects I have worked on, be they large or small, have not been particularly complex.

The contract will state that the contractor's programme must be submitted for approval or acceptance by the employer. Provided the programme contains all elements of the works, reflects a realistic and easily identified critical path, meets the required start and completion dates and is properly resource-loaded, I can see no reason for the contractor's programme not being accepted or approved.

With careful planning of construction activities and resources, the contractor will be able to perform efficiently. Close monitoring of progress and resources, together with early corrective action when required, will prevent or at least minimize delays, budget overruns and potential disputes.

Records

The importance of good record-keeping by all parties, employer and contractor alike, cannot be overstated. Accurate records are essential for monitoring progress, for managing resources and costs (budgets) and for substantiating extension of time, delay-cost and disruption claims.

The Society of Construction Law (SCL) deals quite extensively with record-keeping in Guidance Part B and Appendix B in the second edition of its *Delay and Disruption Protocol*. The Protocol provides excellent guidance to employers and contractors as to the categories of records that should be kept and managed. The categories listed in Appendix B are summarized as follows.

(i) Programme records. These set out, among other things, the contractor's plan for carrying out the works.

(ii) Progress records. These identify the progress of the works at any given time.

(iii) Resource records. These records document the labour, materials and equipment utilized on the works.

(iv) Cost records. These demonstrate the costs incurred in carrying out the works and assist in substantiating amounts claimed in delay and disruption claims.

(v) Correspondence and administration records. These are the written communications relating to the management of the works and contract administration, along with registers, notices, etc.

(vi) Contract and tender documents. These documents are key source documents for establishing entitlement for delay and disruption claims.

It is worth taking note of the sound advice and guidance provided in the SCL Protocol and, if they are adopted by the parties, they will contribute towards a well-managed and conflict-free project.

Performance

The contractor's primary objective for the success of any project must be to *get the job done*. Manage the project well, get the job done efficiently and the project objectives will be realized. The main objectives being

   (i) to execute the works to the highest of standards and
   (ii) to complete the project on time and within budget.

As we discussed earlier in this chapter, to achieve these objectives, the contractor must ensure that adequate and suitably experienced staff and operatives are engaged on the project; that the works are well planned; and that progress, resources and budgets are constantly monitored and managed.

Getting the job done and done well applies equally to the employer. Employers and their representatives have responsibilities (obligations) to perform. Examples of employers' obligations were discussed earlier in this chapter. Performing these obligations in an efficient and timely manner will assist in ensuring a project runs smoothly and will avoid/minimize disputes.

Failure of either party to perform will have an adverse impact on the project and will often lead to blame – and the blame-game inevitably ends in conflict, arbitration and damaged relationships.

THE EMPLOYER'S TAKING OVER OF THE COMPLETED WORKS

Taking over of the completed works by the employer is an important step in any construction project. The taking over process usually involves the following steps.

(i) The contractor issues a notice to the employer that the works are complete and requests the engineer to issue the Taking Over Certificate.

(ii) The engineer will, on behalf of the employer, prepare a list of defective or incomplete works, some of which need to be rectified or completed prior to the Taking Over Certificate being issued. The engineer will also identify any work that can be dealt with after the issue of the certificate.

(iii) When the engineer has issued the Taking Over Certificate, half of the retention money being withheld by the employer will be paid to the contractor.

The Taking Over Certificate is important for three reasons:

(i) it passes responsibility for the care and maintenance of the project from the contractor to the employer;

(ii) the contractor's liability for delay-damages ceases on the date of the certificate; and

(iii) the contractor is paid half the retention money, which can be a significant amount.

The certificate also stipulates the date the defect liability period begins. During the defect liability period, which is usually around twelve months, the contractor is responsible for completing any works (usually minor) that were not complete at the time of taking over, and for remedying defects that become apparent in this period. At the end of the defect liability period, provided the engineer and employer are satisfied that all works have been satisfactorily completed and all financial matters have been agreed, the Final Certificate will be issued, signifying the end of the project works – the end of the contract between the parties.

In the majority of projects I have worked on over the years, the Taking Over Certificate has not been an issue. There have been occasions, however, when the employer has refused to take over the works, often giving spurious reasons (excuses) for not allowing the engineer to issue the Taking Over Certificate. This can create serious problems for the contractor.

- Firstly, the contractor remains responsible for the care and maintenance of the works for an extended period when, by rights, the employer has this responsibility.
- Secondly, the contractor is deprived of the retention money being unreasonably withheld by the employer, adding to the contractor's already troubled cash flow.
- Thirdly, until the certificate is issued, the contractor is exposed to potential delay-damages (liquidated damages) being imposed by the employer.

Such tactics by the employer – tactics to avoid taking responsibility for the project – will invariably lead to disputes and ultimately to arbitration or litigation.

MANAGING PEOPLE

As hard as we may try to prevent disputes, there will always be occasions when we disagree, argue or fall out. It is human nature. Emotion often gets in the way of common sense. What is of prime importance, however, is how we deal with our differences when they surface. How we manage disputes. How we manage people.

People can be a real risk in the construction industry. It is people who dictate whether a project runs smoothly or runs into trouble.

Inexperience, emotion and egos often cloud judgment and are typically why disputes occur. All parties, be they employer, consultant, contractor or subcontractor, should have a policy for disputes to be reviewed objectively and unemotionally by senior management: the people who are not directly involved in day-to-day site management.

Close monitoring of potential disputes by senior management – the decision makers – will give them the opportunity to focus on the issues from a more objective perspective. In my experience, many potential disputes have been quickly resolved when senior management engage in early and constructive dialogue.

An example of disputes being quickly resolved when senior management intervened, was on a project in South-East Asia.

The dispute related to a change in the welding specification for large steel beams. The specification changed from continuous welding to spot-welding. The steelworks in question were undertaken by a subcontractor. The dispute arose over the excessive cost deduction demanded by the contractor's site manager in charge of structures. The subcontractor's site manager believed the cost reduction demanded by the contractor was outrageous and argued there should be no cost saving whatsoever. The two site managers argued for several months without agreement. When it became clear an impasse had been reached, the subcontractor filed for arbitration.

The request for arbitration drew the contractor's senior management's attention to the issue. As the contractor's stand-by mediator, my opinion was sought. It was clear to me that the cost reduction demanded by the contractor's site manager was unreasonable. This issue alone being the incentive for the request for arbitration, I recommended the cost reduction for the spot-welding be withdrawn when parties' senior management met in a final effort to resolve the situation and avoid arbitration. The contractor's senior management recognized their site manager's inflexibility and accepted my recommendation. The dispute was settled within a matter of hours. Arbitration was avoided.

The contractor's site manager complained bitterly about the settlement reached by senior management. Had this matter been left to him, the outcome would have undoubtedly been costly arbitration and a damaged relationship.

The intention of all parties, particularly in the early stages when they are all friends, is to be part of a successful and dispute-free project. The 'honeymoon period' lasts until the works are not progressing as planned, or when the contractor submits its first notice of a claim. This is when the blame-game begins.

When the teething troubles begin, the parties need to have a strategy to deal sensibly with the problems – to manage the situations and the people, to avoid initial differences developing into full-blown disputes. As we observed in the South-East Asia steelwork example, high-level management recognized shortcomings in their staff and focused on the issues, without being influenced or distracted by the people directly involved.

A feature of people that I have observed over the years – a feature that is often the root cause of misunderstanding, disagreement and conflict – is poor communication. I will discuss the importance of communication in the next chapter.

SUMMARY

All parties to a construction project should have the same goal: to execute and complete the works to a high standard, on time and within budget. If these goals, these common objectives, are achieved, disputes will have been prevented, or at least minimized.

The importance of careful planning and thorough preparation cannot be overemphasized. Careful planning by the employer at the conception phase and thorough preparation of tender and project documents by the employer's professional team are essential for the success of any construction project. Careful planning and preparation by the contractor, before digging the first hole in the ground, will contribute enormously to a project being completed on time and within budget.

Each party's rights and obligations under the contract must be clearly understood by all. While being aware of each other's rights is important, the parties should focus on and fully understand each other's and their own obligations. If the parties fulfil their obligations, the project will be a success and disputes are likely to be avoided.

People can be a risk when it comes to conflict in the construction industry. It is people who dictate whether a project runs smoothly or runs into trouble. To ensure a project runs smoothly, we need to manage the people. Many differences of opinion and potential disputes can be quickly resolved when senior management from both parties engage in early and constructive dialogue.

CHAPTER 2

Wise Contract and Claim Management

Listening to and understanding each other's positions together with realistic expectations will invariably lead to sensible solutions.

E ffective contract and claim management are important for maintaining good relationships, for encouraging early and amicable settlement of claims and for avoiding disputes. My thoughts on some important aspects of contract and claim management are summarized in this chapter.

ESTABLISHING THE FACTS

Before entering into dialogue or preparing/submitting claims, the facts must be checked to ensure the claims can be fully supported. The records must be examined and the actual facts established. It is inadvisable to rely on *perceived* facts. A case based on actual (proven) facts will succeed. I will discuss the importance of establishing the facts in more detail in the next chapter, 'Construction Claims: The Investigation'.

COMMUNICATION

Clear and effective communication helps understanding. With understanding comes awareness and knowledge. Poor communication, on the other hand, will lead to misunderstanding and conflict. Effective communication is essential for the success of any project and for preventing disputes. Effective communication requires the parties to talk to each other, to listen, to hear and to understand

– understand what the other party is saying, or is trying to say. If the other party's point of view or reasoning is not understood or heard, it will be difficult, if not impossible, to find a solution.

On the South-East Asian project we discussed in the previous chapter (the project with the steel welding dispute), negotiations between the contractor's procurement department and a potential subcontractor over the terms and conditions of the subcontract agreement had been ongoing for several weeks, without resolution. As the contractor's unofficial mediator, I was invited to attend one of the meetings with the subcontractor in an effort to ascertain what was preventing the parties from reaching agreement.

Within a few minutes of the meeting starting I realized what the problem was – what was preventing the parties from reaching agreement. The subcontractor had been invited by the contractor's senior management to attend the meeting and explain their concerns. During the meeting, whenever the subcontractor's team began voicing their concerns, the contractor's two procurement managers (I will call them the terrible twins) – the two people who had conducted all previous failed negotiations – interrupted, repeating their own positions relating to the terms and conditions over and over again. The terrible twins had no interest in the subcontractor's point of view or concerns. They never gave the subcontractor a chance to be heard. After an hour of this travesty I called for a coffee break. During the break I convinced the procurement senior manager to split the twins and for only me and one of them to return to the meeting. I also asked the manager to insist that the twin accompanying me keep his mouth shut unless and until I asked him to contribute. For the first time in many weeks the subcontractor was heard. I listened to their concerns without interruption. I understood. I offered some explanations and suggested they engage an independent contract consultant to attend the next meeting – to give them reassurance. To give them some comfort. The next meeting, without the twins, was the final meeting. Both sides listened, heard each other's positions, understood and finally agreed. The subcontract was signed a few days later.

Listening and understanding does not necessarily mean agreeing, but listening and understanding will go a long way towards finding solutions.

What I believe in, and have practised throughout my career, is to engage in dialogue in an effort to resolve issues. Talk to each other before entering into written correspondence. I encourage my team to try to reach agreement, or at least an understanding of the other party's position, in face-to-face meetings. Follow up these discussions with an old-fashioned letter or email confirming the outcome of the discussion. This approach invariably helps the parties reach agreement, or at least crystalize their differences without falling out.

The written word, if not carefully structured, can and often does *create* rather than *prevent* disputes.

The correspondents (the authors of the written words) often lose sight of the project objectives and get bogged down with scoring 'ego-points'. This inflames emotions and invites conflict. Points of view can be objectively and calmly expressed without antagonism or insult. Disrespectful correspondence is counterproductive and will not be persuasive. When we are objective and respectful in our correspondence, differences can be quickly and amicably resolved.

TIMELY NOTICES

This is a vitally important contract provision and a requirement that is very often overlooked by contractors. Failing to comply with notice provisions in contracts is the reason why many claims are rejected, particularly when the notice provisions are made a condition precedent to the contractor's rights, such as rights to extensions of time and additional costs. While there may be legal question marks over a party losing its rights for failing to submit the required notices within the stipulated timeframe, it makes good sense for contractors to ensure they comply with these requirements.

Adhering to notice requirements will prevent claims being 'time-barred', whether reasonably or not, and will help avoid disputes. Testing the validity of contract 'time-bar' provisions in arbitration or in the courts is not recommended.

> On a project I worked on in Greece a while ago, the employer rejected the contractor's claim because the notice of claim was submitted *one day* later than the time stipulated in the contract. The matter was referred to the Dispute Adjudication Board (DAB), who, in their decision, applied the express terms of the contract. The contract stated: 'If the claiming party fails to give notice of a claim within the time stipulated in the contract, the contractor shall not be entitled to any extension of time and shall not be entitled to any additional payment.' The DAB ruled in favour of the employer. As unfair as this decision may seem, the DAB was bound to rule in accordance with the express provisions of the contract. Had the contractor complied with the notice provision, which is a requirement in every contract I have worked with, there would have been no time-bar issue and the matter would not have had to be referred to the DAB.

Personally, I am not sure if an arbitral tribunal would have agreed with the DAB's decision, had this issue been referred to arbitration. As an arbitrator, I would have difficulty denying the contractor compensation for legitimate entitlement simply because it missed the notice provision deadline, particularly if the notice was just one day late.

Giving notice of a claim within the strict timeframe stipulated in most contracts is not the only consideration to be taken into account by the contractor. The precise wording of the notice is also important. In some contracts, the contractor must state in the notice the *contractual basis* for making a claim. Failing to strictly comply with the contract regarding such notice provisions can, as we will see in the case discussed below, jeopardize the contractor's claim.

> In a recent case in Hong Kong,* the Hong Kong High Court and Court of Appeal confirmed that the subcontractor (Bauer Hong Kong Limited) had failed to specify in its claim notice the contractual basis on which it later relied, and as such the claim

---

* *Maeda Kensetsu Kogyo Kabushiki Kaisha (also known as Maeda Corporation) and China State Corporation Engineering (Hong Kong) Limited v Bauer Hong Kong Limited* [2019] HKCFI 916; HCCT 4/2018 (9 April 2019).

was rejected. In this case, Bauer encountered difficult (onerous) ground conditions, which required additional excavation works. Bauer gave notice of a claim, citing the additional works as being a variation under the contract. Later, Bauer decided that unforeseen ground conditions would be a more appropriate contract provision to claim under. In the arbitration that followed, Bauer argued its case citing two alternative contract provisions: (i) a variation and (ii) unforeseen ground conditions.

The arbitrator dismissed the variation argument (cited in Bauer's notice) but accepted the unforeseen ground conditions as a valid claim.

The Hong Kong High Court overruled the arbitrator's decision. The Judge's decision was later confirmed by the Hong Kong Court of Appeal. The High Court Judge ruled that in its notice of claim, Bauer failed to state the contractual basis (unforeseen ground conditions) that it later relied on in arbitration.

This case illustrates the consequence of not being fully aware of and adhering to the specific requirements of the contract. The importance of submitting notices of claims within the stated timeframe and ensuring the notices comply with the specific requirements stipulated in the contract cannot be overemphasized. As we have seen, failing to comply with the strict letter of the contract is likely to result in rejected claims, disputes, arbitration and adverse Appeal Court rulings.

REALISTIC EXPECTATIONS

Exaggerated, unrealistic and unsupported claims will result in rejection and, inevitably, disputes. Claims must be based on proven facts and contractors should be realistic in their expectations.

During a lunch in London many years ago I was discussing contractors' claims with a quantity surveyor friend, who at the time was a partner in an established and well-respected international quantity surveying practice. He told me that whenever he was negotiating with a contractor over claim submissions, his opening gambit was to offer the contractor one-third of the amount claimed. His reasoning was that contractors calculate what their claim is really worth and then multiply by three. I responded by suggesting that contractors might be multiplying by three, if indeed they did, because they knew my quantity surveyor friend would divide by three in every case. By tripling the size of their claim, the contractor would usually receive a realistic settlement.

I am not for one moment suggesting contractors adopt the tactic of inflating their claims in the manner described above. If contractors were to apply such tactics, they would lose credibility and their claims will most certainly be rejected.

SUMMARY

To prevent or minimize disputes, all parties must learn to communicate – talk to each other, listen to each other, understand each other's views and positions. Effective communication will go a long way towards preventing misunderstandings, disagreements and conflict.

Late notices of claims, or no notices at all, run the risk of claims being rejected on a 'time bar' basis, whether such rejections are reasonable or not. Time-barring contractors' claims can, and usually does, lead to disputes. To avoid unnecessary disputes, contractors should ensure they submit notices in strict accordance with the contract.

Wise communication and realistic expectations will prevent, or at least minimize, disputes and will go a long way towards finding that sensible solution.

CHAPTER 3

Construction Claims: The Investigation

*An essential ingredient for credible and persuasive claims is prepara-
tion – investigation. A well-prepared claim based on facts, supported
by the evidence, will succeed.*

As suggested in the previous chapter, claims based on the facts
and supported by evidence are likely to succeed, increasing the
prospect of early and amicable settlement.

Before even thinking about putting pen to paper – before starting to
write any part of a claim – a detailed investigation into the key aspects
of the project relating to the proposed claim must be carried out.

The documents and records that need to be thoroughly examined
and fully understood by the claims team during the investigation
stage must, at the very least, include the following.

- The key terms and conditions of the contract.
- The issues (the 'events') giving rise to potential claims. Events
  for which extensions of time and payment of additional costs are
  being pursued by our client.
- The contract programme, updates and progress reports.
- Correspondence relating to the events and each parties' positions
  regarding liability and entitlement.
- Minutes of meetings recording the events that are being considered for the claim.
- Any other records available to support the claim, such as drawings, photographs and, if available videos.
- Previous claim submissions and employer's responses, if these
  exist.

The in-depth investigation of these key components will reveal the strengths and/or weaknesses of each of the events and will help create a strategy for preparing a credible and persuasive claim.

UNDERSTANDING THE CONTRACT

The contract entered into by the contractor will dictate the scope, rights and obligations of the parties. It is essential that the claims team understands the contract. Understanding the contract will help develop a strategy for our client's claim.

There are numerous standard forms of contract used internationally. A few examples I have worked on over the years include

- FIDIC (International Federation of Consulting Engineers) contracts, used in many countries worldwide;
- NEC (New Engineering Contract), used widely in the United Kingdom, also being used internationally;
- JCT (Joint Contracts Tribunal) contracts, used mainly (or only) in the United Kingdom; and
- PPP (Public–Private Partnership) contracts, which are becoming popular worldwide for projects between governments and the private sector.

In addition to the well-known standard forms of contract available, many employers adopt their own 'bespoke' contracts. Some, but not many, bespoke contracts generally follow standard forms, with key terms and conditions modified – at times extensively.

The FIDIC 1999 editions of their contracts, or amended versions thereof, are widely used in the various parts of the world I have worked in. For this reason, my focus throughout this book is on FIDIC contracts and, in particular, on two specific FIDIC 1999 editions; *Conditions of Contract for Construction* (the Red Book) and *Conditions of Contract for Plant and Design-Build* (the Yellow Book). FIDIC has recently published its 2017 editions of the Red Book and Yellow Book. However, as these two contracts are new and, to my knowledge, not yet widely used (if used at all), I have confined my discussions to the FIDIC 1999 editions.

The two types of construction contracts I have worked with in various parts of the world and which I now discuss are build-only contracts and design-build contracts.

Under traditional build-only contracts, with a few possible exceptions (like certain specialist works), the employer is responsible for the entire design of the project and the contractor is to execute the works in accordance with the employer's design. Whereas under a design-build contract, the employer develops its conceptual design and, based on this concept, the contractor prepares the detailed design. Under design-build contracts, the contractor is responsible for both the design and the construction of the project.

Additional costs resulting from variations (changes) instructed under the traditional build-only contract will entitle the contractor to additional payment, with the final values to be agreed between the contractor and employer. If the instructed changes result in delay to the works and the contractor can demonstrate that the contract completion date and key milestone dates were delayed as a result, it is likely an extension of time will be granted by the employer. Justified claims for reimbursement of delay and disruption costs resulting from instructed variations (changes) should also be successful.

Under a design-build contract, however, the contractor will have difficulty in succeeding with claims for any time and cost impact of design changes, the simple reason being that, unless the employer instructs changes to be made to its original concept design, any changes made to the design will be the contractor's responsibility. Additional costs (variation, delay and disruption costs) will be borne by the contractor, and any delay to the completion date (or key milestone dates) will expose the contractor to delay-damages (liquidated damages).

Many construction disputes result from misunderstandings over the parties' responsibilities, particularly under design-build contracts.

Key contract provisions (terms and conditions)

Irrespective of the type of contract in place for the project we are investigating, the importance of knowing the contract – being fully aware of the key terms of that contract – cannot be overemphasized.

Key contract terms and conditions must be thoroughly examined and understood at an early stage of our investigations. Understanding the key conditions (provisions) of the contract may influence the strategy, structure and content of the claim.

A few of the key contract provisions that must be investigated include

- the parties' obligations and responsibilities,
- the programme (start and completion dates and any contract milestones),
- entitlement to compensation for delay and delay-related costs (time and money),
- liability for delay, and
- notice provisions.

The parties' obligations and responsibilities

These have been discussed in some detail in chapter 1. A key consideration when undertaking the investigation is to identify the contractor's obligations and to establish whether or not the contractor has complied with those obligations.

The team also needs to understand the employer's obligations and whether or not the employer has complied. The employer's failure to meet its obligations may strengthen the contractor's claim.

The parties' rights and, in particular, obligations are to be considered for each and every issue/event giving rise to the claim. Establishing whether or not all obligations have been met by the parties will determine the strength or weakness of each issue/event and the claim as a whole. This will also influence the strategy to be adopted for the claim.

The programme

The claims team will spend some time with our client's planners to review the project programme and updates. The programme updates should help identify progress slippages (delays) to individual activities and, more importantly, to the critical path. A careful review of the programme and updates will help the team to establish the cause of delays and to recommend the method of delay analysis to be used to assess the extent of the delay to the contract completion dates.

Entitlement to compensation for delays and delay-costs
Identifying express contract terms that provide compensation for both time *and* money is central to – and goes to the heart of – all construction claims.

Most standard forms of contract are clear when it comes to compensation for delays and to the contractor's entitlement to extensions of time. The FIDIC 1999 suite of contracts state quite clearly the events that will entitle contractors to an extension of time. As an example, subclause 8.4 of the FIDIC 1999 Red Book states:

> The Contractor shall be entitled ... to an extension of the Time for Completion if and to the extent that completion ... is or will be delayed by any of the following causes:
>
>   (a) A Variation ... or other substantial change in the quantity of an item of work included in the Contract,
>
>   (b) A cause of delay giving an entitlement to an extension of time under a Sub-Clause of these Conditions,
>
>   (c) Exceptional adverse climatic conditions,
>
>   (d) Unforeseeable shortages in the availability of personnel or Goods caused by epidemic or government action, or
>
>   (e) Any delay, impediment or prevention caused by or attributable to the Employer, the Employer's personnel, or the Employer's other contractors on the Site.

As stated in subclause 8.4(b), there will be, and usually are, other provisions in the contract that will entitle the contractor to an extension of time, and some that even entitle the contractor to more money. The team will scrutinize the entire contract to locate the terms that provide compensation for both time and money.

Examples of events providing compensation for additional time, costs and sometimes profit that are contained in most standard forms of contracts, and in particular the FIDIC 1999 suite of contracts, include the following.

- Late information: extension of time, costs and profit.
- Denied access/late possession of the site: extension of time, cost and profit.

- Errors in setting out information provided by the employer: extension of time, cost and profit.
- Adverse site conditions: extension of time and cost, but not profit.
- Delays by authorities: extension of time only.
- Engineer's instruction to suspend works: extension of time and cost, but not profit.
- Employer's interference with Tests on Completion: extension of time, costs and profit.
- Changes in the law: extension of time and cost, but not profit.
- Employer's Risk events: extension of time and cost, but not profit.
- Force Majeure events: extension of time only.

Identifying contract provisions that expressly state the contractor's entitlement to additional payment can sometimes prove difficult. The above examples of compensation available for specific events are typical in most standard forms of contract. However, this is not always true for 'bespoke' contracts and amended standard forms of contract.

Most, if not all, contracts I have used require contractors to make reference to specific clauses when claiming compensation for time and/or money. For example, under the FIDIC 1999 Red or Yellow Books, claims for more time and/or money must be made under subclause 20.1. Paragraph 1 of this subclause states:

> If the Contractor considers himself to be entitled to any extension of the Time for Completion and/or any additional payment, ... the Contractor shall give notice to the Engineer, describing the event or circumstance giving rise to the claim. The notice shall be given as soon as practicable, and not later than 28 days after the Contractor became aware, or should have become aware, of the event or circumstance.

That being said, contractors will not be compensated for time or money if the express requirements of subclause 20.1 (or similar clauses in non-FIDIC contracts), have not been strictly complied with. The second paragraph of FIDIC's subclause 20.1 states:

> If the Contractor fails to give notice of a claim within such period of 28 days, the Time for Completion shall not be extended, the

Contractor shall not be entitled to additional payment, and the Employer shall be discharged from all liability in connection with the claim.

If, as expected, our client's contract has the same or similar notice stipulations and consequences as those found in the FIDIC 1999 Red and Yellow Books, and we discover our client has not complied with these express provisions, a strategy for dealing with this potentially serious weakness needs to be carefully considered, discussed and agreed upon with the client.

Delay-damages
Contracts are usually crystal clear when it comes to delay-damages (also known as liquidated damages) to be imposed on the contractor for failing to meet contract completion dates. The claims team will scrutinize the contract provisions relating to delay/liquidated damages, taking careful note of any interim milestones that may carry with them such damages. Awareness of the client's exposure to these damages will be considered by the team and discussed with the client before developing a strategy for the claim.

Liability of the parties for delay and disruption to the works
In my forty-plus years working in the construction dispute business, I am yet to come across a project where one or other party has been correct in every respect. Mistakes are made. No one is perfect.

Our in-depth investigation into the events will undoubtedly unearth weaknesses in our client's case: weaknesses that could damage the case if not dealt with prudently when developing the claim. The strategy for developing the claim will, to a large degree, depend on our client's compliance, or otherwise, with the terms and conditions of the contract.

Mitigation and acceleration
Irrespective of which party is responsible for an event that has resulted in delay and/or additional cost, the contractor has a duty to take measures to reduce the impact of that event.

When, for whatever reason, the works are in delay, contractors are usually asked (or told) to prepare a recovery programme to ensure the contract completion dates are met. This is sometimes interpreted as being an instruction from the employer to accelerate, and as such many contractors believe they will be entitled to compensation for the acceleration costs. The employer usually has a very different interpretation and invariably rejects any acceleration cost claim.

Standard forms of contract are usually clear on the matter of acceleration. Contractors will be entitled to acceleration costs but only when a formal instruction to accelerate is issued by the employer. Mitigation and acceleration obligations are discussed in more detail in chapter 5.

While understanding the terms and conditions of the contract is important, understanding the governing law of the contract is of equal importance. The law may contradict aspects of the contract and may even overrule some contract conditions, particularly if the conditions are onerous. The claims team will consult with the client's in-house legal team, or if no in-house legal team exists it will, with the client's agreement, obtain legal opinion on the governing law of the contract and how the law may influence the strategy for developing the client's claim.

During the investigations, the claims team will be scrutinizing the records and interrogating the client's personnel to find evidence of any mitigation measures taken to help reduce delays and additional costs. The chances of a successful claim will be significantly enhanced if the client can demonstrate they have been vigilant in this regard.

INVESTIGATING THE ISSUES, ESTABLISHING THE FACTS

The issues (events), as known or perceived by the client – that is, the events the client believes delayed or disrupted progress of the works – will have been conveyed to the team during the initial briefing, when the client was telling their story. These events and the client's views relating to compensation will be thoroughly investigated by the team.

Essential features of the events to be explored by the team during the investigations will include

- establishing the cause of each event and the party responsible,
- establishing the impact (effect) of the events on the time or times for completion,
- deciding on the events to be included in the claim,
- establishing the impact of the events on the contract price,
- finding the records (evidence) to support each event, and
- reviewing any previous claim submissions.

Establishing the cause of each event and the party responsible

As I stress throughout this book, the importance of maintaining accurate records cannot be overstated. Accurate records are essential for supporting claims. The records are the evidence to support (justify) each and every aspect of the claim. Without the records – without the evidence – claims will most certainly fail.

Project records play a vital role in establishing the cause of and the party responsible for each event being considered for the claim. The records necessary to establish the cause of and liability for the events include, but are by no means limited, to

(i) the correspondence between the parties relating to the events,
(ii) minutes of progress meetings,
(iii) meetings relating to contractual issues (if they exist),
(iv) regular (usually monthly) progress reports, and
(v) contractor's daily site records (site diary).

An initial review of these records by the team during the investigations will highlight the parties' positions and any disagreements as to the cause of and responsibility for the events. Understanding both parties' positions will help develop a strategy for preparing a credible claim.

Establishing the impact (effect) of the events on the completion dates

In parallel with members of the team scrutinizing the records, the team's delay analyst will be working closely with the client's planners to become familiar with the contract programme, in order to identify the critical path and to examine any changes or updates to the programme. Our delay analyst will also examine the client's progress reports, which will (or should) note any or all delay events and the potential impact of these events on the completion date or dates.

Our delay analyst will also be looking for delays caused by the employer that occurred simultaneously, or in parallel with, delays caused by the contractor. These are referred to in the trade as 'concurrent' delays. Potential or actual concurrent delays identified during the investigation will be taken into account when developing a strategy for the delay element of the claim. Concurrent delays will be discussed further in chapter 5.

From the information gathered during the time spent with the client's planners, together with information gleaned from the review of key records, the delay analyst will prepare a preliminary analysis for each event.

Deciding on the events to be included in the claim

Having scrutinized the relevant documents and completed the preliminary delay analyses, the events that should be focused on (prioritized) in the claim will have been identified. The preliminary analyses will also identify which of the weaker events can, for tactical reasons, be included, but be given lesser priority, as well as which, if any, events could (or should) be excluded altogether.

With regard to the weaker events to be included in the claim, a defence strategy will be developed for future exchanges and settlement negotiations with the employer.

Establishing the impact (effect) on the contract price

Delay-costs
Delays to the project works and, in particular, to the project completion date or dates will inevitably result in additional cost to the contractor. However, delay to the works will not necessarily entitle

the contractor to additional payment. Entitlement to reimbursement of delay-costs will only be forthcoming if the employer agrees to and grants the contractor an extension of time. Delay-costs (often referred to as prolongation costs) will be based on the actual time and duration of each delay event and any extension of time granted by the employer. That being said, not all delay events that entitle the contractor to an extension of time carry with them additional payment.

The investigation into the client's entitlement (or otherwise) to compensation for delay-costs will influence the strategy when developing the cost element of the claim.

Disruption

During the investigations we may discover some events that, while having little or no time impact, might have had a disrupting effect on certain areas of the works, causing unproductive and inefficient working conditions.

The client's daily site reports (site diary) should record any changes to the planned sequence of work – changes that reduced efficiency and productivity. The site reports should note the reasons for any change in planned sequencing and record the extent of work actually achieved against that planned.

On discovering instances of disruptive working conditions, thorough scrutiny of all site records, including the resource-loaded programme, will be carried out by the investigating team. The comparison between planned and actual productivity for the works in question will reveal whether or not to include disruption in the claim.

Variations

If an employer instructs a change in design (or a change in the employer's concept design under a design-build contract) or varies the contractor's scope of work, such change, or variation, will entitle the contractor to be paid for any additional cost incurred in executing the variation. If the change (variation) results in a delay, the contractor will also be entitled to an extension of time. For example, subclause 8.4(a) of the FIDIC Red and Yellow Books states:

The Contractor shall be entitled ... to an extension of the Time for Completion ... by any of the following causes:
(a) a Variation ... or other substantial change in the quantity of an item of work included in the Contract.

The direct cost of executing a change (variation) is seldom, if ever, included in extension of time and delay-cost claims. Variations themselves are usually, if not always, dealt with separately.

Locating the records (evidence) to support the claim

The parties' positions relating to the events and, in particular, to the party responsible for the events will often be found in correspondence between the parties and other records such as minutes of meetings, engineer's instructions, etc.

Persuading the client to provide the records that reflect *both* parties' positions is essential for the team to develop informed opinions on the merits, or otherwise, of the client's case.

On a project in the Middle East, my client, a major international contractor, was convinced their claim had merit and was well supported by documentary evidence.

After our initial investigations, we were disappointed at the lack of records our client's staff were able to locate – records that were essential to support the claims we were in the process of preparing. This problem was reported to our client's senior management, who simply did not believe us. We were instructed to continue preparing their claim and told that the necessary documents, the much-needed records, would be forthcoming 'very soon'. After several more days of liaising with our client's key personnel it was clear that the records we had been promised did not exist.

Another meeting was convened, this time with senior management, site management and the staff responsible for the records all present. It was not a pleasant meeting. The site management finally confessed that the documents needed to support the claim did not actually exist.

> There was a huge loss of face all round, and particularly for the senior management, who had been reporting to head office that they would definitely recover the budget overruns by way of their claim. Our client reluctantly asked us to stop work on the claim.

Reliable records are essential for supporting claims. They provide the evidence to justify/support each and every argument – each and every statement made in the claim. Without the records – without the evidence – the claim will fail.

Reviewing any previous claim submissions

If the client had previously submitted claims, these claims must be reviewed. More importantly, the employer's responses to these claims must be understood, particularly if the claims had been rejected or the employer's valuations were lower than expected by our client.

Understanding reasons for the employer's rejections or unacceptable valuations will help the team develop a strategy for developing the follow-up claim and tactics for future settlement negotiations.

FINDINGS, OPINIONS AND RECOMMENDATIONS

From the findings and opinions formed from the investigations, my team and I will have developed a strategy for a credible and persuasive claim: a claim that should assist the parties in reaching agreement, avoid protracted disputes and, more importantly, avoid having to refer the case to arbitration or the courts.

As soon as the client accepts our findings and recommendations, the planning, preparation and writing of the claim can begin.

Before closing this chapter on the pre-claim investigations, I must mention the difficulties being faced in the international construction industry as a result of the Covid-19 pandemic. Most, if not all, of my contractor clients have been and are still being affected by this pandemic and several of them have asked my opinion on how best to deal with the problem. My research, findings and recommendations relating to the pandemic are discussed in the next chapter.

CHAPTER 4

The Covid-19 Pandemic

'Covid-19 presents an extraordinary challenge, and FIDIC commends all members of the construction community to be focused on the successful delivery of the project before them in a way that sustains the long-term viability of the construction community.'*

I wholly endorse FIDIC's wise advice quoted above. I agree that the construction community, employers, contractors, subcontractors and suppliers alike must remain focused on the successful delivery of their projects and remain dedicated to this goal as long as the impact of the pandemic persists.

Covid-19 created an early dilemma for most contractors, including my clients: should they or should they not close their construction sites? The safety of site operatives, employer's staff and representatives, inspectors and all other personnel visiting their various project sites was a serious consideration.

Remedies and compensation for the businesses and the people directly and indirectly affected by the pandemic has become, and continues to be, a worldwide debate. Within the construction community, the debate invariably leans towards relief for contractors in the form of time and money.

In my research into the impact of the pandemic and possible relief for contractors, I explored how the parties – my clients – can approach this issue to achieve a satisfactory and sensible outcome and, more importantly, avoid conflict.

---

*This is from FIDIC's COVID-19 Guidance Memorandum to Users of FIDIC Standard Forms of Works Contract, April 2020.

The three possibilities for relief I researched were 'Force Majeure', 'Employer's Risks' and 'Change in Law'.*

The first question asked by my clients was: do we close our project sites or continue working? My attempt at answering this question is summarized below.

THE CONTRACTOR'S DILEMMA:
TO CLOSE OR NOT TO CLOSE

In the early part of the year 2020, when the world awoke to the risks of coronavirus and governments began imposing restrictions on all humanity, contractors were faced with a dilemma: to close their construction sites or to continue operating.

When a government directive/order is given to close construction sites, or when instructions to suspend the works are issued by an employer, contractors are obliged to stop work, close the sites and protect the works until the closure or suspension has been lifted. Having complied with such a government order or employer's instruction, the contractor can be, or should be, compensated for any delays, as well as the additional costs incurred as a result of the stoppage or suspension.

In some parts of the world, contractors took the decision to close all their construction sites without having received governmental directives/orders or employers' instructions to do so. I have no doubt that such decisions were not taken lightly by the respective management, and were taken to protect the safety of all personnel working on or visiting their projects. I also have no doubt that, when deciding to close their sites, the decision makers were well aware of the potential contractual and legal consequences they would face.

Any unilateral decision to close a site will undoubtedly place the contractor at risk of being in breach of its contract, resulting in significant damages claims from the employer as well as from its subcontractors and suppliers. Such action will also place the contractor at risk of its contract being terminated.

---

*The form of contract I refer to specifically regarding these three elements is the FIDIC Conditions of Contract for Construction, First Edition 1999 (the FIDIC Red Book).

My clients all recognized this and, without exception, decided to continue operating their construction sites. This posed the next and obvious question, however: what rights will they have under their respective contracts to be compensated for delays and additional costs?

It transpired that every one of my clients' contracts were based on the FIDIC 1999 Red Book – with amendments, naturally. The FIDIC terms relating to Force Majeure, Employer's Risks and Changes in Law were unamended.

Entitlement to more time and money relating to this pandemic is not straightforward. The question I asked myself during my research was: which of these three contract provisions (events or circumstances) can or should the contractor select when making a claim? Initially, I examined the first two: Force Majeure and Employer's Risks. Later in my research I explored the possibility of whether a change in law (or new law) could be relied on for compensation of time and money.

My findings and opinions on the first two contract provisions – Force Majeure and Employer's Risks – are discussed below.

FORCE MAJEURE

Can the coronavirus pandemic be classified as a Force Majeure event?

This pandemic has, in many parts of the world, been referred to as an extraordinary event. Various English dictionaries define the word 'extraordinary' as being synonymous with 'exceptional'.

Subclause 19.1 of the FIDIC 1999 Red Book states:

In this Clause, 'Force Majeure' means an *exceptional* event or circumstance:
  (a) which is beyond a Party's control,
  (b) which such Party could not reasonably have provided against before entering into the Contract,
  (c) which, having arisen, such Party could not reasonably have avoided or overcome, and
  (d) which is not substantially attributable to the other Party. [Emphasis added.]

I believe it is reasonable to conclude from the very nature of the pandemic, from the wording of subclause 19.1 (particularly subparagraphs (a)–(d)), and from the various English dictionary definitions that this pandemic can be classified as a Force Majeure event.

Having established that the coronavirus pandemic can be classified as a Force Majeure event or circumstance, my next question was: under which Force Majeure provision or provisions in the contract will the contractor be entitled to an extension of time *and* payment for additional costs?

Extension of time

Subclause 19.4 of the FIDIC 1999 Red Book is clear:

> If the Contractor is prevented from performing any of his obligations … and suffers delay … by reason of such Force Majeure, the Contractor shall be entitled subject to Sub-Clause 20.1 [Contractor's Claims] to:
>
> (a) an extension of time for any such delay, if completion is or will be delayed, under Sub-Clause 8.4 [Extension of Time for Completion].

Provided the contractor has complied with the contractual pre-requisites for filing a successful claim (such as notice requirements, timely submission of a fully detailed and substantiated claim, etc.) and has demonstrated that the impact of coronavirus caused delay to completion, the contractor will be entitled to an extension of time in accordance with subclause 8.4(d) of the FIDIC Red Book. This subclause states:

> The Contractor shall be entitled … to an extension of the Time for Completion if and to the extent that completion … is or will be delayed by any of the following causes:
>
> (d) Unforeseeable shortages in the availability of personnel or Goods caused by *epidemic* or governmental actions. [Emphasis added.]

Compensation for delay caused by the Covid-19 epidemic (classified by worldwide 'authorities' as a pandemic) seems clear. However, will contractors be compensated for additional costs under this Force Majeure event?

This was the question my clients were most interested in.

Payment for additional costs

With regard to reimbursement of additional cost resulting from a Force Majeure event, the picture is not quite so clear.

Subclause 19.4 of the FIDIC 1999 Red Book states that if a contractor incurs cost by reason of a Force Majeure event or circumstance, it shall be entitled to payment thereof. However, subparagraph 19.4(b) states that the contractor will be entitled to payment of costs, but only 'if the event or circumstance is of a kind described in sub-paragraphs (i) to (iv)* of Sub-Clause 19.1'.

None of the events listed in subparagraphs (i)–(iv) apply to the coronavirus, and as such they will not entitle the contractor to be compensated for costs resulting from the pandemic. However, the second part of subclause 19.1 states that Force Majeure is not limited to the events/circumstances listed in subparagraphs (i)–(v).** The second part of subclause 19.1 states:

> Force Majeure may include, *but is not limited to*, exceptional events or circumstances of the kind listed below [subparagraphs (i)–(v) are those listed below], so long as conditions (a) to (d) above are satisfied. [Emphasis added.]

Interesting, but somewhat confusing – for me at least.

So, as Force Majeure is not limited to the listed events, can my clients successfully argue that they are entitled to payment for

---

*The events listed in 19.1 (i)–(iv) include war, hostilities, etc.; rebellion, terrorism, civil war, etc.; riot, commotion, strike, etc.; and munitions of war, explosive materials, contamination by radioactivity, etc.
** Subparagraph (v) includes natural catastrophes such as earthquakes, hurricanes, typhoons or volcanic activity.

additional costs resulting from this pandemic, a Force Majeure event? This may very well be a reasonable argument, and it is certainly worth considering.

That being said, will the employer agree to payment of such costs? I have my doubts. Will an arbitrator or judge be persuaded? The answer to the latter may well be forthcoming in the not-too-distant future. However, I am of the opinion that if a cost claim for this pandemic has to be referred to arbitration or to the courts for a resolution, which unfortunately may be a reality, a tribunal or judge may not be persuaded. The reason for my skepticism is that I have always understood the core principle of Force Majeure to be that, while the contractor may be entitled to more time, each party should bear their own costs. Tribunals and judges may also have similar views relating to Force Majeure.

EMPLOYER'S RISKS

An alternative to applying Force Majeure as the vehicle for claiming an extension of time *and* payment for costs resulting from the coronavirus is to argue that this pandemic falls under subclause 17.3 of the FIDIC 1999 Red Book: Employer's Risks.

Subclause 17.4 states that if, and to the extent that, any of the risks listed in subclause 17.3 result in loss or damage to the works, and if the contractor suffers delay and incurs additional costs from rectifying such loss or damage, the contractor shall be entitled to

(a) an extension of time for any such delay, if completion is or will be delayed ..., and
(b) payment of any such Cost ...

It is understandable that *loss* or *damage* to the works could result from any of the events listed in subclause 17.3. However, of the events listed in this subclause, only subparagraph (h) could possibly be associated with the coronavirus, and only then if this pandemic can be categorized as a 'force of nature'. Subparagraph 17.3(h) states:

The risks referred to in Sub-Clause 17.4 are:

(h) any operation of the *forces of nature* which is Unforeseeable or against which an experienced contractor could not reasonably have been expected to have taken adequate preventative precautions. [Emphasis added.]

Much has been written about, and argued over, whether Covid-19 is man-made, with some claiming that it originated from a laboratory in China, either intentionally or by accident. Conversely, some scientists reject the theory that the virus was man-made, citing scientific evidence that the virus originated in wildlife. Other scientists advocate that the virus evolved in nature.

Intelligence shared among five nations (the United States, the United Kingdom, Australia, New Zealand and Canada) suggests that it is highly unlikely that the virus was spread as a result of an accident in a laboratory, but rather originated in a Chinese market.*

Notwithstanding the confusion and controversy over whether or not this pandemic can be classified as a force of nature – a controversy that may never be resolved – the forces of nature referred to in subparagraph 17.3(h) are likely to be forces of nature that could cause *loss* or *damage* to the works: forces such as earthquakes, hurricanes, volcanic activity, etc. In reality, the coronavirus does not fall into any of these categories.

Will an employer be persuaded that the coronavirus is a force of nature and therefore an Employer's Risk? I cannot foresee this happening. I also doubt whether a judge or tribunal will be persuaded that the coronavirus can be classified as a force of nature and therefore an employer's risk. In my humble opinion, any such claim is likely to fail.

CHANGES IN LAW (OR NEW LAWS)

Under subclause 13.7 of the FIDIC 1999 Red Book (Adjustments for Changes in Legislation), contractors will be entitled to additional time and cost as a result of any new law or change in law.

---

* Report by CNN's China Reporter Nectar Gan dated 6 May 2020.

During my research into this subject I talked to three lawyer friends who very kindly provided their opinions on whether or not the restrictions imposed by governments for the coronavirus pandemic constitute a change in the law or a new law. Their opinions are summarized below.

My first lawyer friend mentioned that the UK Coronavirus Act of 2020 grants the government delegated authority to issue laws (stay at home, close construction sites, etc.). The question asked by this lawyer was: does this delegated authority mean a change in law (or new law)? He was doubtful. In his opinion, the risks should be shared in line with the *force majeure* principle: more time but no money.

The second lawyer did not see the measures imposed by governments to be changes in law, or fitting within the change in law option under the contract. If claiming additional payment under a change in law provision, contractors need to be specific as to the particular law and its relevance.

My third lawyer friend was of the opinion that the measures imposed by governments or local authorities may indeed relate to some change in law. The problem was identifying that law. If their claim is to succeed, contractors will need to be specific about which law.

Subclause 13.7(b) provides cost relief for contractors in the event of new laws or changes in the law. However, it is not certain (it is still being debated among prominent international lawyers) that the measures imposed by governments do constitute new laws or changes in the law. If, as I have been led to believe, some governments have in fact decreed a new law or changes in existing laws, and if contractors suffer delay and incur additional costs as a result thereof, they will be entitled to compensation for both time and money. That being said, and as noted by my lawyer friends, for claims to succeed under changes in law or new laws, contractors must identify the specific law in question.

CONCLUSIONS AND RECOMMENDATIONS

Establishing entitlement to additional costs resulting from the pandemic is proving to be a challenge.

Many clients and consultants I have discussed this situation with feel it is unfair for contractors alone to be burdened with these costs. I too believe that this is unfair. I also believe that employers – and indeed arbitrators and judges – may also think it unfair, and I am certain they will all be sympathetic. Unfortunately, sympathy will not necessarily convert into money.

Contractors will be awarded more time to complete the works without penalty (delay-damages). Small compensation I know, but not necessarily insignificant. A well-presented and properly supported extension of time claim under Force Majeure will, I believe, succeed.

Whether or not to claim payment for costs under any of the three categories discussed in this chapter should be carefully considered by the contractor's decision makers. Cost claims will almost certainly be rejected by employers, and the claims may have to be referred to arbitration or the courts to have any chance of succeeding.

Both the employer and the contractor will have suffered, incurring additional costs. A reasonable, and I believe sensible, approach would be for the parties to bear their own costs. This approach is what I have always understood and endorsed as a core principle of Force Majeure.

Claiming an extension of time 'only' under Force Majeure for this pandemic seems prudent and is likely to result in early amicable settlement. More importantly, such an approach will ensure that the important relationships between the parties are maintained.

CHAPTER 5

Preparing Claims to Persuade a Tribunal

When submitting a claim, our goal must be to persuade the other party that the claim has merit and is good enough to persuade an arbitral tribunal. Such a claim should be an incentive to persuade the parties to seek and find a sensible solution.

During a recent in-house training session I presented, the subject being 'Preparing claims to persuade a tribunal', one of the participants, a senior member of a consultancy firm, said 'we are only preparing a claim for our client' and asked: 'Why do we need to provide evidence to support the claim? Surely, this [providing evidence] can wait until the case goes to arbitration?'

When tasked with preparing a claim for contractors, our clients, we need to ask ourselves what the purpose of the claim is, what we are aiming to achieve and what the goal is.

The goal must surely be to prepare a claim that will succeed. To develop the means – the tool – for our clients to reach a realistic and satisfactory outcome. To avoid protracted disputes. The goal must surely be to avoid having to refer the claim to arbitration for a resolution.

KEY CONSIDERATIONS WHEN WRITING A PERSUASIVE CLAIM

Despite what I have just said about the goal being to avoid having to resort to arbitration, when we begin writing the claim, we should be thinking ahead – thinking about the *possibility* that the claim will eventually be referred to arbitration for a final resolution. With this thought in mind, we must ensure the claim is good enough to persuade a tribunal.

If the claim is good enough to persuade a tribunal, the employer should recognize this and be encouraged to negotiate and agree a fair settlement – to find that sensible solution.

To persuade the employer and, if necessary, a tribunal, the claim must at the *very least* be clear, objective, comprehensive but concise, credible and realistic.

Clear

The claim narrative (the claim report) must be clear and easy for the reader to understand. Problems can arise when the narrative is complicated or so boring that the reader loses interest. This is particularly true when the claim report – the story – is full of technical and legal jargon. When writing the claim, use language that will be readily understood by all.

Objective

The tone of the claim report is important. Objective arguments and respectful phraseology can be persuasive. A confrontational approach to claims – blaming, alleging incompetence or accusations of bad faith – will incite conflict and endanger relationships. Adopting a confrontational approach will make it difficult, if not impossible, to achieve a sensible outcome.

Comprehensive but concise

Keep the claim report short. Describe the issues (the events) as concisely as possible. By the time the decision makers have read the claim report, they should have a clear understanding of the contractor's position with regard to all the events and what it is the contractor is asking for. The claim report must be written in such a way as to hold the reader's attention from beginning to end.

Detailed chronologies, analyses, calculations, etc., if included in the claim report, will sidetrack the reader and run the risk of the reader (the employer) losing interest. Such details can and should be included in appendices to the claim report.

Credible

For claims to succeed, they must be based on *actual* facts and not *perceived* facts. The *actual* facts must be supported by the evidence – by documents, the records. Many claims I have witnessed have failed because those preparing the claims have not bothered to examine the records or to find the evidence to support their arguments. They often include statements based on what they understood or believed to be true: statements/arguments based on what they 'perceive' to be the facts. Claims that cannot be supported will most certainly fail.

Realistic

The claims must be realistic, particularly claims for additional costs. Exaggerated claims, which are usually obvious to an employer and to a tribunal, will fail. Unrealistic, exaggerated and unsupported claims will be rejected by employers, and this, more often than not, leads to disputes.

STRUCTURE AND OUTLINE CONTENT OF THE CLAIM

The results of the investigations (discussed in chapter 3) will dictate the strategy, content and tone of our client's claim. From the investigations we will have established which of the events should be focused on and which events should be downplayed or even excluded.

Before we put pen to paper – before we begin writing the claim – I insist on preparing a simple structure and outline content and agree this with our client. The structure and outline content usually takes the shape set out in the following example.

The Claim Narrative (Claim Report)

Executive summary. Concise overview of the claim.

Introduction. Brief summaries of the project, the type of contract, the parties, the scope of work, the contract price, the commencement and completion dates, a list of the events, and the time and money being claimed.

The events. Description of each event prioritized from the strongest event to the events having the least strength or impact.

Delays. Brief explanation of method of delay analysis; a summary of delays resulting from the individual events; and overall delay to contract completion dates.

Mitigation. Measures taken by the contractor to reduce (mitigate) the impact of the events in question.

Delay-costs. Explanation of 'heads of claim' and summary of amounts claimed under each head of claim.

Disruption. Explanation of disruption to the works caused by certain events; summary of disruption costs.

Changes/variations. Rarely included in delay/disruption claims.

Summary. Compensation claimed (relief sought): time and money.

All supporting documents will be included in the appendices to the claim report. The appendices, discussed a little later in this chapter, are what I refer to as the claim's 'engine room'. They contain the analyses, cost calculations, event chronologies and records – the evidence needed to support the claim.

Having concluded the investigations and agreed the strategy, structure and outline content with our client, writing the claim can finally begin.

THE CLAIM REPORT

As discussed earlier in this chapter, the claim report (the claim narrative) should be comprehensive but concise, and it should be kept simple, be easy to understand and be written in a way that will hold the reader's attention from the beginning to the end. Most important of all, the report must be credible, respectful, based on the facts and supported by the evidence – the records.

We will now look at each section of the claim report, as set out in the structure and outline content example provided above.

Executive summary

As the title implies, this is a summary of the contractor's claim aimed at the employer's executive – its senior management: the decision makers. The executive summary is vitally important, as it is intended to persuade the employer's executive that the claim has merit and should be taken seriously.

The executive summary should provide a concise account of the contractor's claim – summarizing the entire claim document (the full story) in just a few pages. The reason for such an abbreviated version of the claim is simple: most senior executives have neither the time nor the desire to read the entire story – the full claim. These executives need to be convinced, in very few words, that the claim is credible.

Condensing the entire claim into a few pages and ensuring the key elements are covered and are clear can be a challenge. Only a few of my colleagues in the construction claims business are able to write clear, concise and effective executive summaries.

In one particular claim I was asked to review, the executive summary ran to fifty-five pages – longer than many entire claim reports. The claim report itself was in excess of 900 pages, and the appendices, which included the complete set of contract documents, filled several lever-arch files.

It was not surprising to learn that the claim was immediately rejected by the employer and returned, unread, to the contractor. The employer's executive management had not even bothered to begin reading the executive summary, let alone the 900-page report.

The employer advised the contractor that it has never been impressed or persuaded by the size of a claim – by the number of pages or the weight of the document. The employer suggested that if the contractor expected its claim to be taken seriously, it had to rethink its strategy and submit a more sensible claim.

It is vitally important that the employer's decision makers are encouraged to read the executive summary. If, after reading the summary, the executives are persuaded that the claim is credible – that it has merit – the chances of a sensible solution will be greatly enhanced.

The executive summary should be the final section of the claim to be written, the simple and obvious reason being that it summarizes all aspects of the fully developed claim, highlighting the cause and effect of each event, the extension of time and additional costs the contractor believes it is entitled to and is claiming.

Introduction

The introduction section of the claim report should also be short. The introduction is merely a formality and should only note the following key elements of the project and the contract:

- a brief description of the project;
- the names of the parties;
- a brief summary of the scope of the construction works;
- the start and completion dates, and any milestone dates;
- the contract price;
- a list of events giving rise to the claim; and
- the additional time and money being claimed.

The events

This is one of the most important sections of a contractor's claim. It tells the story. In this section we identify the events giving rise to the claim. Here we explain the 'what', 'when' and 'why' of the event or events.

From the initial investigations we will have identified what happened, when the event happened and why it happened (the cause). Having established the 'what', 'when' and 'why', we should be able to demonstrate the impact (effect) of the event or events on the project completion date or dates and on the contract price. Having identified the cause and effect of each event and the effect on the project as a whole, we need to establish the specific terms of the contract that will entitle our client to be compensated by way of an extension of time and payment for the additional costs incurred.

Quotes from contract documents

The contract terms and conditions (clauses) that provide for entitlement to compensation will be identified and referred to in this section of the claim report. It can be helpful to quote short extracts from the specific clauses that support our client's entitlement. Having said that, in general, my preference is to quote extracts from the contract in an appendix to the claim report, particularly if the quotes are lengthy, and make *reference* to the relevant contract terms in the 'events' section of the claim report.

I have on occasion witnessed most, if not all, contract terms and conditions being quoted verbatim, either in the introduction section of the claim report or within the section dealing with the events. This I find astonishing: it is totally unnecessary and rather irritating. The employer and its representatives will have the complete set of contract documents. Only relevant and short extracts need be provided – extracts that identify and emphasize entitlement for more time and money.

Prioritizing the events

If, as so often happens, there are several events to be considered for inclusion in the claim, a strategy needs to be agreed with the client as to the priority of the events: a strategy that will set out the events in the sequence that will be most persuasive. Applying this strategy, we can focus on those events that are the strongest – the events that are clearly the employer's responsibility and will entitle the contractor to fair compensation: the events that can be supported by the evidence. The weaker, less credible events should be relegated to the bottom of the list. The thinking behind this strategy is that once the stronger events have been reviewed, the employer may be persuaded as to the overall credibility of the claim. Once the employer is persuaded the contractor does indeed have a credible claim and is entitled to some compensation, it only remains for the parties to negotiate and agree the level of that compensation.

Structuring a persuasive argument
Some of the key factors to be considered when telling the story –
when writing the claim report – were discussed earlier in this chapter.
These important factors are true for all parts of a claim and essential
when arguing our case for each event. They apply equally to the
strongest and weakest events.

Let us remind ourselves of some of these essential factors and
how they should be applied when presenting a persuasive argument
– when preparing a persuasive claim.

Base the arguments on the facts – facts that can be supported
by the evidence: the records. If an event cannot be supported by the
records, then it would be wise not to pursue that particular event.
Not only will events that cannot be supported fail, they may jeopardize the entire claim.

The description of the events must be easily understood. Use
clear, simple terminology when describing the events – particularly
the technical aspects of the events. Avoid using technical jargon. Do
not overwhelm the employer's senior management – the decision
makers – with unnecessarily complex technicalities.

Keep the description of the event – the 'what', 'when' and 'why'
– short. Describe the event as concisely as possible. We must hold
the employer's attention from the beginning to the end. We need to
convince the employer that our claim has merit.

A picture is worth a thousand words
A well-used adage this may be, but it is certainly true, and especially
when it comes to supporting claims. For almost all construction
projects I have worked on, contractors keep extensive and detailed
photographic records of the works. Some contractors also make
video recordings.

These photographic and video records are extremely useful for
our claims. They illustrate what actually happened and highlight
some of the difficulties encountered by the contractor. While photographs and videos alone are not in themselves sufficient, when
they are combined with a well-structured argument (the thousand
words), they will greatly enhance the prospects of a successful
claim.

An important exercise during the initial investigations is to examine our clients' photographic records in the hope of finding good examples to use in the claim. The following story is an example of how using just a few photographs can contribute to a successful claim.

The project included a large pump station, which was central to the project and critical to completing the works within the stipulated time. The contract was 'build-only', the employer having full responsibility for the design.

Having excavated the majority of the 'pit' for the pump station, the contractor discovered a large cave-like opening. The opening can be seen in the far corner of the pit in the photograph below.

This photograph was used when describing what the contractor discovered while excavation works were being carried out.* The claim report explained that part of the excavated pit suddenly collapsed, exposing the hole. The report explained that all

* While the resolution of the photographs might not be very high – as was the case in the claim report from which the three photos in this section are borrowed – what is important is that the pictures are worth those 'thousand words'. The same principle applies to the bar charts used a little later in the chapter.

works in close proximity to the hole were suspended, the area surrounding the hole was cordoned off for safety reasons, and the employer was immediately notified. The event (discovering the cave) and the action taken by the contractor were supported by the contractor's daily site reports.

On examining the hole, the employer instructed the contractor to cease all works within the pump station and the surrounding areas until a geotechnical expert could investigate the situation. The employer's instruction to cease works became an important document as evidence to support the contractor's eventual claim. On advice, the contractor issued a notice of potential delay to the employer. It transpired that the hole was in fact a large water-filled cave, as illustrated in the next photograph, which was also included in the claim to highlight the size of the cave.

Based on the geotechnical expert's recommendation, subsurface explorations were carried out to all areas of the project works. The employer instructed the contractor to suspend all works on the project until the investigations were complete and the geotechnical expert had submitted its report. This instruction also became an important piece of evidence to support the contractor's eventual claim.

Again on advice, the contractor issued a follow-up notice to the employer of the delay being caused by the suspension of all works and advised the employer of the potential additional cost.

At this stage, my team and I were engaged by the contractor to begin investigations towards preparing a claim on their behalf.

The geotechnical expert concluded that the cave needed to be pumped dry and filled with weak-mix concrete. The employer instructed the contractor to carry out this task as a variation to the project works. Once the cave had been filled, the employer confirmed that the suspension had been lifted and that works could resume. As can be seen from the date on the pump station pit photograph below (4 October 2008) and the date on the first photograph, when the cave was discovered (27 October 2007), the project was delayed for approximately one year.

Based on our investigations and findings, we prepared our client's claim. Our client's entitlement to a one-year extension of time, delay-costs and 'variation' costs for pumping and filling the cave with concrete were all supported by the documented evidence, which included the photographs. The claim, drafted on the principles described in my book, persuaded the employer that the contractor was entitled to more time and money.

> After intensive, but mostly amicable, negotiations, all claims were agreed to the satisfaction of both the employer and our client, the contractor.

The above example shows how a claim based on a combination of careful research/investigation, well-structured factual argument illustrated with photographs and supported by the records (the evidence) will usually succeed.

Movies
Another useful aid to persuading employers to seriously consider contractors claims is to convert detailed claim submissions into short, well-scripted, expertly produced and professionally directed videos (movies).

My first encounter with the creation of claim movies was when I was working for a construction company in the Middle East. One of my duties as 'contract coordinator' for the company was to oversee three extension of time and delay cost claims. While these claims were partially successful on an interim award basis, the company's head office was becoming impatient and demanding early closure. The story of making my first claim movie is told below.

> One day, when visiting the project, the chairman of the company's board of directors asked to see me. He had a plan. I was to prepare a what he termed a 'global' claim. It took me a while to understand what he really intended: I was to summarize the three claim submissions into a very short document, no more than twenty pages he suggested, summarizing our precarious cash-flow position and requesting a 'high-level' meeting in an effort to agree on a sizable advance payment and to early settlement of all three claims.
>
> Around the same time, I met the owner of a claim consultancy firm who introduced me to his pride and joy: the construction claims movie. I pitched the idea of a movie to the board members and they liked the idea, so my fourteen-page 'global' claim document became the script for a movie.
>
> I then met the movie maker Rob Valenta. Rob is a movie technician extraordinaire. We made a good team. I was the

scriptwriter, producer and director; Rob was the genius movie maker. Within six weeks, a ten-minute movie was produced that was highly acclaimed by the board of directors.

The closing scene of the movie – and indeed the theme throughout it, as well as being the general theme of my book – concluded with the following statement: 'A solution can be found by engaging in constructive dialogue.'

Both the movie and the script were well received by the employer, who even suggested the contractor's entire claim should be presented in the form of a movie – it would save them all a considerable amount of reading.

The movie and the concisely worded 'script' were so well received that, for a moment, Rob Valenta and yours truly contemplated relocating to Hollywood to further our movie-making careers. I look forward to producing further 'award-winning' movies that will contribute to the early settlement of many more construction claims.

Delays – impact on completion

Having established that the events to be included in the contractor's claim were caused by, or were the responsibility of, the employer, we must demonstrate – prove – that the events, individually and collectively, actually delayed the project works.

The contractor's programme will be the basis on which delays to the works are identified and against which the impact of such delays will be measured.

Proving the events delayed completion and the extent of the delay
A delay analysis, while essential, will not on its own prove that the delays were caused by or were the responsibility of the employer or that the contractor is entitled to an extension of time. To persuade the employer, and perhaps later a tribunal, the contractor has to

(i) prove that delays were caused by the employer;
(ii) show when and for how long the delay-events happened; and
(iii) demonstrate that the delays impacted on the critical path, thereby delaying the completion date/dates, and by how much.

The proof – the evidence – will (should) have been found during the investigations, when scrutinizing the contractor's site records – the daily site records, or site diaries as I knew them in my younger days. These site records, if maintained correctly, will record the dates on which any delay event occurred and when it ended, what exactly the event was and what or who caused the event. The site records should also detail which areas of the works were affected.

In addition to scrutinizing the contractor's site records, our team will have researched all correspondence, minutes of meetings, progress reports, registers of drawings, procurement records, etc., relating to each of the events. From this research our team will have prepared a chronology for each event. These chronologies, which should be limited to pertinent issues relating to the particular events, will be used by the analysts to develop their analyses.

Methods of delay analysis
According to the Society of Construction Law (SCL) Protocol,* there are six commonly used methods of delay analysis:

   (i) impacted as-planned analysis,
  (ii) time impact analysis,
 (iii) time slice analysis,
  (iv) as-planned versus as-built analysis,
   (v) retrospective longest path analysis and
  (vi) collapsed as-built analysis.

I have experienced only three of these: 'impacted as-planned', 'time impact analysis' and 'as-planned versus as-built'. Neither the impacted as-planned analysis nor the time impact analysis proved particularly successful, with both being open to criticism and rejection by employers.

The impacted as-planned analysis was criticized – rightly in my opinion – for being theoretical and unable to provide proof that the delay events actually impacted on the completion date. The one occasion an impacted as-planned method was successful was when

---

* SCL Delay and Disruption Protocol (2nd edition, February 2017).

the employer delayed initial handover of the site for the contractor to commence work. The contractor was awarded an interim extension of time for the one-month delay for late access. The contract programme simply shifted by a month.

Time impact analysis is a complex, time-consuming and costly method of analysis, and, in my experience, provides too many openings for the employer to question and criticize.

The as-planned versus as-built analysis has potential, as it seems to be based on actual impact on progress. If supported by the records, this method can be less open to criticism.

I do not profess to be a planner or delay analyst in any way, but in my earlier days, before computers took over the world, I prepared programmes for several medium to large construction projects in the United Kingdom. I also prepared delay (extension of time) claims on a few of these projects. Both the programmes and the analyses were prepared by hand on A3 sheets of paper, using a drawing board and a T-square (my tools of the trade). The delay analyses for the extension of time claims were uncomplicated and easy to understand. Without exception, all delay claims I prepared, using those tools of the trade alongside a commonsense approach, were settled amicably.

I still keep my drawing board and T-square in the futile hope that common sense will one day prevail and that computer analyses will be consigned to the past. I have no doubt that arbitral tribunals and judges would welcome a more commonsense approach to delay analyses.

Presenting the arguments for extension of time
Presenting convincing arguments for extension of time is the most important and, at times, the most difficult section of the claim. It is in this section that we must prove to the employer (and maybe later, to the tribunal) that the contractor is entitled to an extension of time.

The principles for presenting persuasive arguments in this section of the claim are the same as for all other sections of the claim. Without shame, I yet again highlight here some key principles outlined earlier in the chapter.

- The arguments must be clear and easily understood.
- The story being told must be concise. It must hold the reader's attention at all times.
- The arguments must be credible (believable) and supported by the evidence: the proof that the employer-delays impacted on the critical path activities, and actually delayed completion.

Describing the impact of the events should be confined to an accurate but brief account of the effect of the employer-delay events, summarizing the extent of delay caused by each individual event and the combined effect of all events on the project completion date or dates. This narrative will also state the relevant terms of the contract that entitle our client to an extension of time and, from the analyses, the amount of additional time being claimed.

Earlier in this chapter I quoted the well-worn adage that a picture is worth a thousand words, and I provided examples of the kind of photographs that can support a claim. The same principle applies to the impact of delay events, but instead of photographs we use simple bar charts to illustrate the delay events against the original programme and the impact of these events on the project completion date. An example of a simple bar chart is provided below.

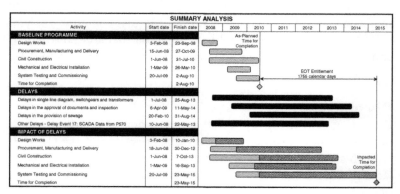

When providing these bar charts, it must be explained in our narrative that the delays and the impact thereof are summaries only, and are derived from the detailed delay analyses contained in the appendix to the claim report.

It is important to remember that every statement and argument made in the claim report *must* be cross-referenced to the supporting documents provided in the various appendices. This applies equally to the bar chart illustrations.

A summary of the delay events, the impact on the contract completion date or dates, the extension of time sought and the contractual basis for the extension of time should be set out at the end of this subsection of the claim report.

Being granted an extension of time is important for two reasons: (i) it will relieve the contractor of liability for delay-damages (liquidated damages) and (ii) it establishes the contractor's right to compensation for delay-costs.

Concurrent delay

Concurrent delay is a topic that has been debated for quite some time within the construction community, but no consensus has yet been reached.

There are differing schools of thought when it comes to contractors' compensation in the event of concurrent delay – that is, when delays caused by the employer run in parallel with delays caused by the contractor.

Some differing views still being debated include that

(i) the contractor is entitled to extension of time for the overall delay but is not entitled to recover any delay-costs;

(ii) both time and delay-costs should be apportioned pro rata to the delays caused by the parties;

(iii) there is full entitlement to both time and delay cost if the employer's delay occurred before the contractor's delay but no compensation if the contractor's delay occurred before the employer's delay; and

(iv) compensation for both time and delay-costs will depend on which of the delay events were more dominant.

Apportioning entitlement to time and cost seems fair but, in practical terms, difficult to establish, particularly with regard to delay-costs. In my view, the SCL Protocol sums up this difficulty well:

Where Employer Delay to Completion and Contractor Delay to Completion are concurrent and, as a result of that delay the Contractor incurs additional costs, then the Contractor should only recover compensation if it is able to separate the additional cost caused by the Employer Delay from those caused by the Contractor Delay.

The difficulty is separating (apportioning) both time and cost caused by the employer and by the contractor. This is where the contractor's records come into play. In my experience, contractors' records are seldom, if ever, sufficiently detailed to fulfil the task of separating delay and delay-costs in the event of concurrent delay.

I must confess that I do not have an answer to the concurrent delay debate. It seems fair to apportion both time and cost if the apportionment can be assessed or analyzed, but that is a difficult, if not impossible, task.

Mitigation

Most, if not all, contracts I have worked with – be they standard or 'bespoke' forms of contract – are clear regarding contractors' obligations to carry out the works diligently and to complete the works by the date or dates stated in the contract. For example, the FIDIC 1999 Red and Yellow Books state in subclause 8.1 that the contractor 'shall proceed with the Works with due diligence and without delay'. Then, in subclause 8.2, it says that the contractor 'shall complete the whole of the Works, and each Section (if any), within the Time for Completion for the Works or Section (as the case may be)'.

The contractor should, from the very start of the works, be monitoring actual progress against the planned activities set out in the contract programme. Close monitoring of progress and resources together with early corrective action when required will prevent, or at least minimize, delays, cost/budget overruns and potential disputes.

The contractor's monitoring and reporting progress is an express obligation under the FIDIC 1999 Red and Yellow Books. So too is the contractor's obligation to notify the employer of the measures being taken or to be taken to overcome any delay to the progress of the works. In other words, the contractor is obliged to mitigate delays to the works for which it (the contractor) may be responsible. Subclause 4.21 (Progress Reports), paragraph (h) of the FIDIC 1999 Red and Yellow Books states that within each of the contractor's monthly progress reports, the contractor shall include 'comparisons of actual and planned progress ... and the measures being (or to be) adopted to overcome delays'.

All contracts I have encountered require contractors to issue 'recovery programmes' when progress is inconsistent with the original programme or any revised programme. Under the FIDIC contracts, the employer (or engineer) can instruct the contractor to submit

> a revised programme and supporting report describing the revised methods which the Contractor proposes to adopt in order to expedite progress and complete within the Time for Completion ... *at the risk and cost of the Contractor.*[*] [Emphasis added.]

I understand and agree that it is perfectly natural for the contractor to rectify (mitigate) its own shortcomings and to do so at its own cost. However, I do not believe the contractor has an obligation to mitigate delays and bear the costs resulting from events for which the employer is responsible.

The SCL Protocol suggests that contractors have a general duty to mitigate the effect of an event for which the employer is responsible. However, the Protocol states that, subject to the express wording of the contract, the contractor's duty to mitigate does not extend to requiring the contractor to add extra resources or to work outside its planned working hours (i.e. the contractor is not obliged to mitigate employer risk events at its own cost).

While it would be prudent for the contractor to do its best to minimize the impact of an employer's risk event, this should not be an obligation in my opinion. That being said, it will help – and in my experience it has helped – contractors' claims to be settled amicably when the contractor can demonstrate that mitigation measures were

---

[*] FIDIC Red and Yellow Books, subclause 8.6 (Rates of Progress).

taken, naturally at no cost to the contractor, to reduce the impact of delays caused by the employer.

Any such mitigation measures will need to be proved. The mitigation measures need to be clearly demonstrated in the detailed delay analyses provided in an appendix to the claim report. The effect of the mitigation measures can be illustrated in a simple bar chart diagram. Such an illustration – something similar to the example below – should help convince the employer that the contractor's mitigation measures have been effective.

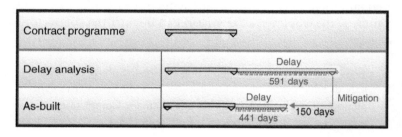

Mitigation or acceleration?

Disputes often arise when contractors claim reimbursement for taking acceleration measures to recover employer-delays without receiving an instruction from the employer to accelerate. Most, if not all, contracts I have worked with have specific provisions relating to acceleration. Contractors can recover the cost of accelerating the works but only if the employer issues a formal instruction to accelerate.

When contractors accelerate the works to recover delays for which the employer is responsible without a written instruction to accelerate, they often claim the cost of these acceleration measures under 'constructive acceleration', which has become a common term. This approach rarely works and usually results in a dispute.

When delays are caused by, or are the responsibility of, the employer, agreement should be reached between the contractor and the employer as to the remedies to mitigate such delays. If such remedies result in the contractor incurring additional cost, the employer must agree to reimburse the contractor. Such agreement should be reached before any mitigation measures are implemented. If the employer refuses to compensate the contractor, then contractors should not, in my opinion, be obliged to undertake mitigation measures.

Delay-costs

Contractor's additional cost claims are, more often than not, the most difficult for the parties to agree on. Additional cost claims are usually the main cause of disputes, and they often lead to arbitration. While delay (or prolongation) cost claims should be relatively straightforward to justify and reach agreement on, they are often rejected, or the claimed amount is drastically reduced by the employer.

When developing the delay-cost element of the claim, we must persuade our clients not to claim delay-costs for events that do not entitle the contractor to additional payment, including, for example, exceptionally adverse weather, delays caused by local authorities, epidemics such as Covid-19 (discussed in chapter 4) and Force Majeure. Delay-cost claims should be based solely on the delays that, under the contract, will allow the contractor to recover these costs.

Delay cost categories (heads of claim)

It is not unusual to find several categories of delay-costs – categories commonly referred to as 'heads of claim' – appearing in contractors' claims. 'Heads of claim' are, in my view, divided into two general groups: (i) credible or potentially strong and (ii) less credible, weak or impossible.

The 'heads of claim' that may fall into the 'credible' category, will be

  (i) direct costs;
 (ii) indirect costs, often referred to as site overheads;
(iii) unrecovered head office overheads;
 (iv) extended insurances and bank guarantees/bonds; and
  (v) finance/interest charges.

Less credible (sometimes impossible) 'heads of claim' that are occasionally included in contractors' claims are

(i) loss of profit and
(ii) claim preparation cost.

Direct costs
Direct costs relate to site labour, plant and equipment. Delays may not per se result in additional labour, plant or equipment costs unless the delays disrupt the efficient productivity of the works. Delays that disrupt productivity are common occurrences, and the cost impact is usually claimed in a separate disruption section of the claim. Disruption is discussed a little later in this chapter.

Indirect costs or site overhead costs
The indirect or site overhead costs generally relate to site establishment and operating costs. Examples of these indirect (or site overhead) costs include the following.

- Project staff based on the site: staff such as site management, supervision, procurement, planning, contract administration, ancillary/support staff, etc.
- Site offices: setting up, operation, maintenance and removal.
- Dining facilities: setting up, maintenance and removal.
- Labour camps: setting up, maintenance and removal.
- Transportation.
- Site security, health and safety.

The most significant site overhead costs will be the site-based project staff.
    Under most contracts I have worked with, compensation for delay-costs is based on actual and proven costs. This principle is supported by the SCL Protocol,* which notes that

---

* Guidance Part B: Guidance on Core Principles, Item 20: Basis of calculation of compensation for prolongation.

compensation for prolongation … is based on the actual additional cost incurred by the Contractor. The objective is to put the Contractor in the same financial position it would have been if the Employer Risk Event had not occurred.

An important consideration when calculating site overhead costs – and one that is often overlooked or misunderstood by some contractors – is that site overhead costs should be based on the costs being incurred at the time (or times) the employer-delays actually occurred. These costs should be available from the contractor's site records.

This principle is supported by the SCL Protocol,* which states:

> Once established that compensation for prolongation is due, the evaluation of the sum is made by reference to the period when the effect of the Employer Risk Event was felt, not by reference to the extended period at the end of the contract.

The delay analysts must therefore ensure that their analyses clearly identify the periods in which the employer-delays occurred and impacted progress of the works. It is these periods on which the site overhead costs should be calculated.

The contractor and/or the employer have, on numerous occasions, adopted a simpler approach to calculate site overhead costs: they have used the 'extended preliminaries' approach. This process identifies the items priced in the 'preliminaries' section of the contract price (bills of quantities) that are time related. The prices/amounts in the preliminaries section are converted into a daily price (or rate) based on the contract period. This daily rate is then applied to the period (in days) of delay, to arrive at the cost claimed. This easy-way-out calculation is often favoured because (i) it saves having to calculate/evaluate the actual cost, which can be time consuming; and (b) the contractor's records may not be sufficient to support an actual cost claim. In my experience, employers

---

* Guidance Part B: Guidance on Core Principles, Item 21: Period for evaluation of compensation.

sometimes prefer the 'extended preliminaries' method because the resultant cost is, more often than not, considerably less than when using the 'actual cost' method.

Unrecovered head office overheads and profit
The approach often used to justify claims for head office overhead costs and profit, and one I have used myself, is that the contractor invests in resources to undertake various projects, which collectively earn revenue. The revenue earned includes an element of head office overhead costs and profit. When projects are completed on time, the resources are released to earn revenue for the contractor from new projects. When projects are delayed, however, the resources are tied to that project and are not able to generate additional revenue during the delay period, resulting in unrecovered overhead costs, as well as lost profit.

To successfully claim unrecovered head office overheads (and loss of profit) the contractor must be able to demonstrate – to prove – there was a revenue-earning project (or projects) available during the delay period but that had to be turned down or postponed because its resources were tied up in the delayed project. The difficulty is in demonstrating (proving) such lost opportunities.

It is quite common for contractors to claim unrecovered head office overheads and loss of profit using a formula rather than trying to prove actual overhead costs, with the two most-used formulae being the Hudson and Emden formulae. The Hudson formula uses the contractor's overhead and profit mark-up included in the contract price, whereas the Emden formula uses the contractor's proven overhead obtained from audited accounts.

When using either the Hudson formula or the Emden formula, the contractor needs to make adjustments for any overhead and profit already recovered in variations. This adjustment is necessary to ensure there is no danger of 'double-recovery'.*

---

* A third formula called the Eichleay formula is, as I understand it, used primarily in the United States. I have never used this formula and nor have I noticed it being used in any claims I have reviewed or evaluated.

Insurances and bank guarantees/bonds
It is a contractual obligation for the contractors to provide insurances and guarantees/bonds, and they must be valid for the duration of the project. If the project is delayed, these insurances and guarantees must be extended, usually at an additional cost. If the extended periods result from delays for which the employer is responsible, the contractor will be entitled to recover these additional costs. Naturally, the contractor will have to provide evidence of the additional premiums and bank charges.

Finance/interest claims
I have included this head of claim in the 'credible' category because many commentators, including the SCL Protocol,[*] consider that claims for finance/interest charges for unpaid additional costs are justified. I agree with this principle, but in my experience these claims are usually rejected by employers and are often withdrawn by contractors during settlement negotiations as a trade-off for other, more significant, 'heads of claim'.

Loss of profit
Most contracts I have worked with do not provide for contractors to recover loss of profit as part of delay-costs. Having said that, under FIDIC 1999 forms of contract, certain employer-responsible events allow recovery of time, cost *and* profit. The difficulty faced by contractors when claiming profit in their delay cost claims is isolating the events and delay durations for the specific events that allow for recovery of profit. This is a complex exercise and is rarely carried out by contractors.

When advising my clients I make it clear that recovery of profit in their claims is likely to be rejected by employers. I suggest my clients agree to withdraw the profit part of their claim as a trade-off during settlement negotiations – an approach that often works.

---

[*] Guidance Part C: Other Financial Heads of Claim, Item 1: Claims for payment of interest.

Claim preparation cost

This head of claim has rarely, in my experience, been used, and on the few occasions contractors have insisted on claiming for the cost of preparing the claim, it has failed. In my opinion, it does not even feature as a trade-off in settlement negotiations. According to some commentators, including the SCL Protocol,* the contractor will not be entitled to the cost of preparing the claim unless the contractor can show it incurred this cost as a result of the employer's unreasonable actions (or inaction) when dealing with the contractor's claim. While this claim might succeed under the circumstances identified in the SCL Protocol, it will do so only if the disputed claim is decided on in arbitration or litigation. I always try to persuade my clients against including this head of claim: if pursued, it will more than likely end in rejection, dispute and damaged relationships.

Summary: delay-costs

The delay cost claims must, as for all sections of the claim, be credible, realistic and supported by the evidence. The money claim is always the most sensitive part of any claim and is the element most scrutinized by the employer and arbitral tribunals.

For each and every cost being claimed, the contractor must include details of the cost calculations and provide evidence of the actual costs. Such evidence will include staff names and salaries, copies of audited accounts, etc. When such evidence is of a sensitive or confidential nature, the contractor should invite the employer to visit their offices to undertake its own (confidential) audit. This is usually sufficient to satisfy employers. The calculation sheets and documented evidence (if not confidential) will be included in the appendices to the claim report.

A strategy for settlement negotiations should be decided by the contractor when preparing the cost element of the claim. The focus of the cost claim should be on the 'credible heads of claim', with the less credible heads included but withdrawn in a trade-off during the negotiations. This has worked in the vast majority of claims

---

* Guidance Part C: Other Financial Heads of Claim, Item 3: Claim preparation costs.

and negotiations I have been party to. When applying this strategy, claims have been settled, and disputes, arbitration or litigation have been avoided.

In closing, my advice to all my contractor clients is not to play the game my London quantity surveyor friend played, and divide contractors' cost claims by three because (he claimed) he believed all contractors calculate what they need and treble this in their claim. I have never experienced this game being played in the fifty-plus years I have been in the construction business. Contractors have certainly exaggerated claims that I have seen – all too often, in fact – but exaggerated claims will fail. They will fail simply because they cannot be supported. They will be rejected. This in turn will lead to disputes, arbitration or litigation.

Disruption

Disruption can be described as interruption or disturbance to a contractor's activities, causing less efficient working conditions and lower productivity than planned. As a result of the inefficient and unproductive working conditions, the cost of carrying out specific works will be greater than was planned and priced for. In such circumstances, when disruption is caused by employer-risk events, contractors will be entitled to – and will certainly claim for – the additional costs resulting therefrom.

Proving and calculating disruption costs are historically difficult tasks, as accurate and detailed records are essential for successful disruption claims. Records of planned resources (labour, plant and equipment) for each specific activity of the project works – and, in particular, records of the actual work achieved for these specific activities – are rare or do not exist at all. However, when such records do exist, contractors will have a better-than-even chance of a successful disruption claim.

Various methods are used to calculate disruption costs, but none of them can be said to be accurate or scientific. Having said that, the construction industry itself cannot be classified as an accurate science: not when pricing the works at tender stage, nor when planning the durations of the many work activities, nor when establishing the labour and equipment needed to execute the works. These are all 'estimates' based on experience and productivity factors established from previous projects.

Of the various methods used for evaluating disruption costs, I have selected two for discussion:

(i) the 'measured mile' method and
(ii) the relatively new 'system dynamics' method.

The measured mile
The measured mile is, in my experience, the most used method for demonstrating disruption.

This method compares productivity achieved for parts (areas) of the works impacted by a disrupting event with the productivity achieved in non-affected areas, where the areas of works being compared are the same or similar in nature. In addition, the productivity achieved in the non-affected areas should be the same as, or similar to, the productivity factors used in the contract programme and in the contractor's price.

On a project in the Middle East, the contractor experienced severe disruption to major excavation works. The subsurface data for existing live services provided by the employer at the tender stage, which formed part of the contract documents, did not identify the extent of live services in existence in several areas of the project.

Excavation works had to be executed in a piecemeal fashion, working around the 'live' services, slowing productivity significantly. Many of the resources assigned to these excavation works had to be relocated to other areas of the works, creating organizational problems for the contractor. The disruption also delayed the overall project.

The contractor submitted claims for

(i) an extension of time for the delay to the project;
(ii) additional payment for the unproductive working conditions;

(iii) additional payment for the delay-related costs for the extended contract period; and

(iv) additional payment for resources that were idle for extended periods of time while the contractor reorganized its sequence of working.

The employer's initial response to the contractor's claims was that the accuracy of the site data provided at tender stage was the contractor's risk and responsibility, and therefore all claims relating to delay, disruption and idle-time were rejected.

Naturally, the contractor disputed the employer's position, and arguments between the parties continued for several months. The employer eventually acknowledged responsibility for the pretender site data and a sensible agreement was finally reached on compensation for time and money, including reasonable disruption costs.

What eventually persuaded the employer to agree on the value of the disruption claim was that the contractor was able to demonstrate the following key elements of the 'measured mile' method for demonstrating and evaluating disruption.

(i) The productivity factors for bulk excavation works included in the contractor's resource-loaded programme and in the contract price were realistic.

(ii) The productivity achieved in the non-affected areas of bulk excavation were, as supported by the contractor's records, similar to the productivity factors in the resource-loaded programme and in the contract price.

(iii) Based on well-maintained records, the contractor was able to demonstrate the extent of lost production in the affected areas and to accurately value the disruption cost.

The system dynamics method

The system dynamics method of quantifying disruption, as it has been explained to me, uses two simulation models to calculate loss of productivity.

The first simulation is calibrated to produce an 'as-built' situation, recording the historical performance of the project, inclusive of unplanned events. Once the 'as-built' simulation has been developed, a second 'but-for' simulation is run, removing the direct impact of the unplanned events. The difference between these two simulations records the disruption caused by the unplanned (employer-risk) events.

The SCL Delay and Disruption Protocol* has acknowledged the system dynamics modelling method in its 'project-specific' studies** as a recognized method of disruption analysis, but it notes the following challenges:

(i) the accuracy and completeness of the source input data and the quality and availability of project records;
(ii) the reasonableness of the analyst's judgements in establishing the model; and
(iii) the transparency of the analytical process carried out by the specialist software.

The Protocol concludes by saying that, given these challenges together with the complexity and cost involved, system dynamics is not as commonly used as other methods of calculating loss of productivity. That said, I am reliably informed that there has been an arbitration case where the system dynamics method for calculating disruption costs was successful. The story, as I have been told it, is outlined below.

In a recent construction dispute, the employer rejected the contractor's disruption claim in its entirety, counter-claiming a significant sum in damages from the contractor. The contractor had acknowledged an element of blame for the disruption in its claim and, with the aid of the system dynamics analysis, the

---

* Second edition, February 2017.
** Guidance Part B: Guidance on Core Principles.

contractor concluded that the employer was responsible for 60% of the disruption, while the contractor accepted 40% liability.

The case proceeded to arbitration. The employer maintained its position that the contractor was entirely to blame for the disruption and sought relief for damages. The contractor argued – as it had in its claim and during the unsuccessful settlement negotiations – that it accepted part of the blame but not 100% of it, as proven by the results of the analysis undertaken in the system dynamics simulation.

The tribunal awarded the contractor the 60% of the disruption costs claimed.

I wonder if the tribunal's decision was based on its understanding of the system dynamics simulation and the conclusions of the analysis or whether, in the absence of any persuasive argument from the respondent (the employer), the tribunal was persuaded by the contractor's acceptance of a fair portion of responsibility for the disruption, irrespective of any formula used in arriving at the apportionment of blame.

I have read the literature and attended several discussions/explanations on the workings of this system, and while I understand the principle, I am unable to fully fathom the computerized simulation and the results that emerge from what I call the 'magic box'.

As an arbitrator, I believe that had I been on the tribunal in the above case, I would have based my decision on the 60%/40% distribution of responsibility claimed by the contractor as being reasonable. I would not have based my findings on a computerized simulation model I did not understand.

Changes/variations

Claims relating to variations or changes to the project works are seldom, if ever, included in contractor's delay and/or disruption claims. All contract documents I have worked with contain clear mechanisms and procedures for evaluating changes/variations. While the pricing of changes/variations often leads to disagreement and disputes between employers and contractors, this matter has not been dealt with in my book. Changes/variations have been the subject of many books written by experts who are, without exception, far better versed in this subject than I ever wish to be.

Summary

A summary of the relief sought by the contractor is essential. The summary identifies at a glance what compensation or relief the contractor is requesting. The summary will state

   (i) the extension of time claimed for each milestone (if applicable) and the contract completion date,
  (ii) the relief from delay (liquidated) damages for the delay,
 (iii) the amount of delay-costs being claimed for each 'head of claim',
 (iv) the disruption costs claimed, and
  (v) the total of all costs claimed.

THE APPENDICES

The appendices to the claim report are, as I mentioned before, the claim's engine room. This is where the detailed delay analyses, cost calculations, documentary evidence and detailed chronologies are located. It is the backup to the claim report, and as such it is vitally important that the appendices provide accurate details and factual evidence to support every aspect of the contractor's claim.

A brief outline of typical appendices is provided below.

Contract particulars
Quoted extracts from selected contract terms relevant to the events, highlighting rights and entitlements.

Delay events
A detailed chronology for each delay event, noting the cause, responsibility and timing of each event. The chronologies will be developed from careful scrutiny of the records – particularly the correspondence

– and will highlight any disagreements between the parties as to the cause of and responsibility for the events.

Delay analyses
The delay analysis methodology (or methodologies) used in the claim will be explained in the opening paragraphs of this appendix, stating the reasons why the respective methodologies have been used.

This appendix will also contain the actual delay analyses and results for each event, as well as a consolidated analysis identifying the overall time impact of the delay events on the contract completion date and milestones (if milestones are applicable). The chronologies are useful, if not essential, when developing the delay analyses.

Cost records/calculations
All supporting records and cost calculations for delay, disruption and any other costs being claimed will be found in this appendix. A few examples of the evidence needed to support the costs being claimed include (i) copies of the contractor's records confirming salaries, (ii) head office overhead and profit mark-up included in the tender (contract price), and (iii) extracts from the contractor's audited accounts.

Documents
The evidence. As emphasized more than a few times in my book, every statement and every argument in the claim must be supported by documentary evidence. Examples of this evidence include progress reports, employer's instructions, minutes of meetings, relevant correspondence and photographs.

SUMMARY: PREPARING CLAIMS TO PERSUADE A TRIBUNAL

From the detailed investigations we have undertaken, we will have established most of the facts and located the evidence to support all aspects of the claim. Before starting to write the claim report, we will prepare a structure and outline content, which will be explained to, and agreed with, our client.

For the claim report (the narrative) to be persuasive, it must at the very least fulfill the following criteria.

- It must be clear and easy to understand, particularly by the employer's executives and, later, if necessary, by a tribunal or judge.
- It must be concise. The report must hold the reader's attention at all times. When the report is unnecessarily lengthy, the reader will be easily distracted, or might become bored. Our message can be lost.
- It should be kept simple. Do not overwhelm the reader with complex descriptions or technical jargon. It will be difficult, if not impossible, to hold the reader's attention. As for a long-drawn-out narrative, the reader is likely to lose interest rather quickly.
- It must be credible – believable. All statements and arguments in the report must be supported by the evidence.
- It must be objective and respectful. The tone of the claim report is important. Objective arguments and respectful phraseology can be persuasive. On the other hand, a confrontational approach – blaming, alleging incompetence or accusations of bad faith – will incite conflict and endanger relationships.
- Finally, it must be realistic. The claims must be based on actual and not perceived facts. Cost claims must not be exaggerated: this will jeopardize credibility and be an obstacle for settlement.

The appendices to the claim report are where the details, analyses, calculations and documentary evidence are located. Appendices need to be complete and must fully support every aspect of the contractor's claim.

As stated at the beginning of the chapter, our goal when preparing a claim must be to persuade the other party (the employer) that the claim has merit and to have a claim that is good enough to persuade a tribunal. Such a claim should be an incentive for the employer to seek and find that sensible solution.

CHAPTER 6

Seeking That Sensible Solution: Negotiation

The most economical and the most satisfactory way of resolving differences, of reaching sensible solutions, is by negotiation.

As quoted above, the most satisfactory way to resolve differences – and in particular contractor's claims – is through negotiation. My focus in this chapter is on examining negotiation techniques: techniques that can help parties find that sensible solution.

AFTER THE CLAIM: NEGOTIATION

If the contractor's (my client's) claim is well prepared, easy to understand, based on facts and supported by the evidence, then the chances of reaching a fair and sensible settlement with the employer will be reasonably high. Having said that, there will be occasions when settlement is not reached. In my own experience, one of the main reasons for agreement not being achieved through negotiation has been the approach and, at times, the attitude of the individuals representing one or other party during the negotiations.

On one particular project in the Middle East, for which my team prepared an extension of time and delay cost claim for our client, the claim was prepared exactly as I have described in this book. First we undertook a thorough investigation; then we developed a claim that was good enough to persuade a tribunal; and, most importantly, we provided the records – the evidence – to support each and every argument, each and every statement.

The employer's representatives rejected the claim on the grounds that we had not provided sufficient evidence to support the claim.

It was clear during several 'negotiations' that followed that the employer's representatives had not seen – or had simply ignored – the documents (the evidence) they insisted were missing. Despite directing these individuals to each and every document appended to the claim report – documents they insisted were missing – they continued writing letters repeating their demands for information to support our claim: information they had had in their possession for several months.

These tactics continued without any possibility of settlement. My client finally lost patience and gave notice of their intention to refer the matter to arbitration. This action alerted the employer, who arranged to meet our client at high level, without their representatives present. Agreement was reached soon thereafter, and the notice of referral to arbitration was withdrawn.

While similar tactics by employers' representatives might be experienced from time to time, this is not necessarily common practice. Fortunately, in my experience, the vast majority of employers and their representatives review contractors' claims on their merits and, more often than not, make fair evaluations, which invariably leads to agreement.

In most projects I have been involved with, the initial settlement negotiations take place between the contractor's site management and the employer's representatives – between the people familiar with the day-to-day activities of the project. This can be, and often is, problematic, as the people directly involved usually take a personal interest in the project, so they are sensitive and can be defensive, particularly when trying to establish the party responsible for the events giving rise to the claims. Emotions, attitudes, egos and blame can get in the way of objectivity and common sense. These emotions can, and often do, contribute to unresolved negotiations.

If the initial negotiations fail to achieve agreement, the next stage in the negotiation process should, and usually does, result in agreement. This stage involves senior management of both parties seeking to find sensible solutions and avoid third-party dispute resolution procedures such as arbitration.

PREPARING FOR THE NEGOTIATIONS

Having submitted our client's claim, and in anticipation of the employer's response, it is imperative to prepare our client for the forthcoming negotiations.

The claim will have been prepared in such a way that, if no agreement is reached with the employer, a tribunal will be persuaded. As we discussed earlier, such a claim should also persuade the employer. Having said that, nothing should be taken for granted. We need to have a strategy to persuade the employer that our claim has merit and that a solution should be found during the negotiations.

Preparation before negotiations is essential. Having worked with our client through the investigation stage and then during the development of the claim, our client will (should) fully understand the strengths and potential weaknesses of their case. Our client must be persuaded to have realistic expectations and be prepared to accept an outcome they can live with, even if it was not the outcome they were aiming for.

Something I always prepare before entering into negotiations is a realistic 'high–low' settlement range: a range between the best-case scenario, or the best that can be expected, and the lowest deal my client is prepared to accept. If the other side is willing to settle within my client's expectation range, agreement will be reached.

The techniques for successful negotiations – if they are understood and applied during negotiations, whether at initial stage negotiations or later, when senior management get involved – usually result in amicable settlement.

NEGOTIATION TECHNIQUES

The most natural and cost-effective way of resolving differences and disputes is through dialogue: by negotiation. The two negotiation techniques I have encountered over the years are techniques known in the trade as 'positional bargaining' and 'principled negotiation'.

Positional bargaining

The positional bargaining technique, or tactic, is used when the only goal for the negotiator is winning. Getting everything they want – 100% of it. The position of the negotiator (or party) is entrenched; no compromise is offered. Positional bargaining is often used when the negotiator is not concerned about future relationships. This frequently results in mistrust, prolonged conflict and broken relationships. These tactics seldom work.

A claim I prepared for a client in Greece had been referred to arbitration. I was invited to participate in pre-arbitration negotiations. Each party was represented by legal counsel. The lawyer negotiating on behalf of the other party adopted positional bargaining tactics and it was clear he had only one goal in mind: to destroy our case.

For two hours the lawyer strutted around the room like a horny rooster, letting us all know just how good a negotiator he was. Several times he pointed to all the commendations he had received, which were framed and hanging on every wall of the meeting room so everyone could see just how great he was. At no time during his two-hour rant did he give our lawyer or my client any chance of stating our position.

When it became obvious that we would not be given a chance to present our case, we advised the other party and their cocky lawyer that there was no point in continuing the negotiation, and we left. I noticed when we bid farewell to the other party that they seemed a little embarrassed. A good sign, I thought.

A week later the other party's managing director called my client and asked for a meeting – without any lawyers present. All issues were resolved amicably at that meeting. The positional bargaining tactics used (and used badly) by this particular lawyer failed.

By adopting a more cooperative, problem-solving approach, the parties can avoid the inefficient gamesmanship of positional bargaining. Negotiations can succeed when the negotiators – the parties

– adopt what is known as 'principled negotiation': making an effort to reach an outcome that meets both party's interests.

Principled negotiation

What I have found useful, and mostly successful, in negotiations is to apply simple mediation techniques, with the most effective of these techniques being to listen.

When preparing for the negotiations, we will (or should) have established what aspects of our case are important to us. We also need to realize and accept our own weaknesses.

When sitting around the negotiating table, we must make an effort to listen. We must try to understand the other side's concerns and the reasons behind those concerns – to understand what is important to them that may not be that important to us. We must also ensure the other party is aware that we are listening and that we understand their concerns and interests. Understanding does not necessarily mean agreeing, but it is the first step in reaching agreement.

We should try to establish not only what is in our best interest, but also the interests of the other party (the employer) – not forgetting the interests of the project. We must avoid getting bogged down in our positions. As we have seen when positional bargaining tactics are applied, fixating on positions is often the reason why negotiations fail.

Establishing objective criteria and standards is key to successful negotiations. We should focus on events that have undisputed entitlement and are supported by hard evidence – events that will be accepted by the employer as objective, legitimate and practical.

Our client should be prepared to withdraw claims that are unsupported. Withdrawing such claims can open the door to early settlement.

Compromise is often seen as a dirty word when it comes to negotiations, but if a sensible solution is to be found, compromise may be necessary and it can work, particularly if the end result sits within the realistic settlement range prepared before entering into the negotiations.

SUMMARY

The most natural and cost-effective way of finding sensible solutions is through constructive dialogue: through negotiation. If negotiations are to succeed, careful preparation, realistic expectations and the desire to reach agreement are essential.

If both parties, or at least one of them, adopt some of the principled negotiation techniques outlined above, the chances are that the parties will find that sensible solution. On the other hand, if either party – or, indeed, if both parties – are entrenched in their positions and use positional bargaining techniques, negotiations are likely to fail.

CHAPTER 7

When Negotiations Fail: Alternative Dispute Resolution

*Although Alternative Dispute Resolution processes vary in their style and content, they have a common objective: to bring about the resolution of a dispute without resort to arbitration or litigation.**

In many construction contracts, the first step to be taken by the parties after site-level and/or mid-level management negotiations have failed is for the top management of each party – usually chairmen, chief executive officers or equivalent – to try and settle the issues. In my experience, this high-level dialogue usually works.

If these high-level negotiations fail, the next step in the process is third-party-assisted dispute resolution, known in the trade as 'alternative dispute resolution', or ADR.

Several forms of ADR are used for resolving construction disputes, but the two I am familiar with, and the ones discussed in this chapter, are dispute boards and mediation. I also briefly mention statutory adjudication.

DISPUTE BOARDS

There are two types of dispute boards: dispute resolution boards (DRBs) and dispute adjudication boards (DABs). The difference between them is that DRBs recommend solutions whereas DABs adjudicate and render binding decisions.

---

* This quote comes from Julian Bailey's *Construction Law* (3rd edition): Volume III, Chapter 23, 'Dispute resolution'.

DABs are more commonly used in international construction projects. They are certainly more common in the parts of the world I have worked in – South-East Asia, Europe and the Middle East. As DRBs are not frequently used in international construction projects (apart from in Australia, where DRBs are used in some states[*]), I will focus my discussion on DABs.

Dispute Adjudication Boards

All FIDIC contracts that I have had sight of use DABs. The FIDIC contracts stipulate that disputes shall be *adjudicated* by a DAB. There are two types of DABs: full-time, or 'standing', dispute boards; and 'ad hoc' dispute boards. Under the FIDIC 1999 Red Book,[**] standing DABs are used, whereas under the FIDIC 1999 Yellow and Silver Books,[***] ad hoc DABs are used. Single-member or three-member dispute boards can be appointed for both standing and ad hoc dispute boards.

Ad Hoc Dispute Adjudication Boards
Ad hoc DABs are appointed after a dispute has arisen. The single-member or three-member DAB is appointed by consensus of the parties. Unless otherwise agreed to by the parties, the DAB will expire once it has given its decision.

The function of an ad hoc DAB is merely to adjudicate. At no time is there any interaction between the DAB and the parties. As we will see next, under standing DABs, such interaction can help prevent differences developing into full-blown disputes.

Standing Dispute Adjudication Boards
Standing DABs are more commonly used in the FIDIC contracts I have worked with. In some cases where the FIDIC 1999 Yellow Book was being used and in which ad hoc DABs were stated, the contracts were amended, replacing ad hoc with standing DABs.

---

[*] Lucy Goldsmith and Andrew Stephenson, Corrs Chambers Westgarth, Sydney, Australia.
[**] FIDIC 1999 Red Book: Conditions of Contract for Construction.
[***] FIDIC 1999 Yellow Book: Conditions of Contract for Plant and Design-Build.
FIDIC 1999 Silver Book: Conditions of Contract for EPC/Turnkey Projects.

The reason for replacing ad hoc with standing DABs is, I believe (or would like to believe), that standing DABs take an active role in a project. The DAB members visit the project site regularly: usually every 2–3 months depending on the magnitude and complexity of the project. During these site visits, the DAB members will meet with the parties to ascertain progress and to check if any difficulties have arisen. If difficulties, disagreements or even disputes exist at any time, the parties may, by agreement, approach the DAB for an opinion.*

If the parties are unable to resolve their differences or disputes, either between themselves or after receiving the DAB's opinion and guidance, the dispute may be referred to the DAB for a decision.

If neither party challenges the DAB's decision within a specified time, the decision shall become final and binding on the parties. However, if either party challenges the DAB's decision and the parties are unable to settle the dispute after further negotiations, the dispute may finally be settled by arbitration.

Standing Dispute Adjudication Boards: to help parties resolve differences or to merely adjudicate – to be judge and jury?

I am firmly of the belief that the main function of standing DABs should be to help prevent differences between parties developing into full-blown disputes. The most effective way to realize this objective is, in my opinion, for the dispute board members to guide the parties towards finding sensible solutions – to help the parties settle their differences quickly and efficiently. In other words, the dispute board members should try to instill some common sense into the minds of the parties.

> I attended a dispute board conference in Turkey several years ago. During one of the workshops, the following question was asked: What is the real function of DABs? To help the parties resolve their differences or merely to adjudicate – to be judge and jury?

---

*Requests for a DAB's opinion can be implemented under FIDIC 1999 Red Book at subclause 20.2, paragraph 7.

This question created some lively discussion, particularly when I suggested that DAB members should apply mediation-type techniques when trying to help the parties resolve their differences. My suggestion caused some booing and jeering from a minority of the delegates, who obviously considered mediation a dirty word. I was not deterred and pressed my point. By advocating the use of mediation techniques, I was not suggesting that mediation should replace DABs. My point was, and still is, that such techniques can be useful for DABs when helping parties resolve their differences.

At the conclusion of the workshop, the consensus was that a fundamental objective of dispute boards was indeed dispute resolution: to use a common-sense approach to achieve the real function, or objective, of the DAB – to help the parties find that sensible solution.

If the function of standing DABs was simply to adjudicate when parties were unable or unwilling to resolve disputes among themselves, then the very existence of, or requirement for, standing DABs would, in my opinion, be questionable. If it was indeed the case, parties should choose ad hoc DABs to simply adjudicate.

FIDIC has recognized the importance of dispute avoidance in their second editions of the FIDIC Red and Yellow Books: they have changed the title from Dispute Adjudication Boards to Dispute Avoidance/Adjudication Boards (DAABs),* encouraging the members of the DAAB to help the parties resolve their differences. While similar provisions are available in the FIDIC 1999 suite of contracts, the new title formalizes the requirement for dispute boards to actively assist parties to find that sensible solution.

MEDIATION

Who says mediation resolves disputes? It damn well doesn't. It works ever so well at getting rid of disputes, but don't go telling me it *resolves* them. To get a resolution is to get an impartial, imposed

---

*FIDIC Red and Yellow Books (2nd edition, 2017), clause 21: Disputes and Arbitration.

and binding decision about the facts, allegations, evidence and law by a judge or an arbitrator. The mediator says that if, instead of a resolution, you are attracted to getting rid of the dispute, then give up the idea of seeking a decision about your rights under the contract.

The punchline is rather like a good joke, taking the brain by surprise: abandon your legal rights – embrace commercial convenience instead.

When we mediate, we are buying into a different adventure.*

I smile whenever I read these words of wisdom from Tony Bingham. Although what Tony says may seem to some people to be more than a little unorthodox, to me what he says makes perfect sense. In my rather limited experience as a mediator, I have observed that for mediation to work, both parties must be seeking, or at least hoping for, a resolution – if not on the day of the mediation hearing/meeting itself, then soon thereafter. Should one of the parties not wish to 'embrace commercial convenience' – if either or both of the parties is not prepared to make any sort of compromise – then mediation will be a waste of everyone's time.

Mediation is, and should always be, a voluntary process for seeking and finding resolution. Mediation being a prerequisite to arbitration has become fairly common in many construction contracts. There also seems to be a growing desire in the United Kingdom for mediation to be mandatory before parties can refer their dispute to the courts.

The problem I have with mandatory mediation is that one party, or sometimes both parties, will simply go through the motions of mediating without any intention of settling, particularly if they want 'their day in court'.

The mediation process

Mediation is a process led by an independent or neutral person: the mediator. The mediator actively helps the parties negotiate towards settlement of their differences or disputes. Mediation will give the

---

* These are selected extracts from an article by Tony Bingham titled 'Blow your rights', which appeared in *Building Magazine* on 9 May 2003.

parties' decision makers the opportunity, perhaps for the first time, to hear the other side's arguments and to understand the strengths and, more importantly, weaknesses of their own case and that of the other side.

There are two distinct approaches to mediation: facilitative and evaluative.

Facilitative mediation

In a facilitative mediation, the mediator does not give opinions or pass judgement on the position of either party. The mediator attempts to help the parties, in Tony Bingham's words, 'embrace commercial convenience' – to help the parties find a resolution they can comfortably live with. The mediation hearing or meeting usually takes the following form.

Opening joint session with all parties present
During this session both parties will be invited to make oral presentations, summarizing their respective positions and arguments. The mediator may allow the parties to exchange further arguments or request clarification from the other side. The mediator may also seek his/her own clarification from the parties during the opening session.

Private sessions (caucus meetings)
During these sessions/meetings, the mediator may help each party

- by identifying issues of differing value – one party may place little value on something that may be considered important to the other party and vice versa;
- by reality testing – holding up a 'mirror' to the party and asking how realistic their position or argument is;
- by being a sounding board – allowing the parties to vent their feelings, test their arguments and help prioritize their aspirations; and
- by helping the parties shape proposals that meet their needs and those of the other party.

While the mediator is meeting separately (privately) with each party, the other party will (should) engage in some soul searching and honest discussion among themselves.

Mix of private and joint sessions
The mediator will decide on the number of private sessions and their durations, and he/she may alternate them with further joint meetings in an attempt to narrow down the differences.

Final joint session: getting to settlement
Through the various joint and private sessions, the mediator will have helped the parties reach agreement on all their differences.

The mediator will help draft the settlement agreement, although the agreement will normally be drafted jointly by the parties or their respective lawyers. The agreement will be signed by those authorized to sign for each party and will usually be witnessed by the mediator.

Any forward-thinking mediator will have ensured appropriate drinks are available to celebrate the settlement agreement.

Evaluative mediation

In the evaluative approach, the mediator will be asked to give opinions on the merits of each party's position on the issues and on their case as a whole. The opinions given by the mediator may result in the parties reaching agreement during the mediation hearing.

The mediation 'day' may follow the same or a similar pattern to the facilitative mediation outlined above, with a joint session to start, followed by a mix of private and joint sessions. If, by the end of the hearing, the parties have reached agreement on all issues, a settlement agreement will be drawn up and signed during the final session and the celebrations will follow.

If agreement has not been reached on any or all of the matters in dispute, the mediator may be asked to submit a written opinion on the merits or otherwise of each party's case. Agreement is often reached soon after the parties have received the mediator's written opinion.

STATUTORY ADJUDICATION

I have not personally experienced adjudication under the United Kingdom's Construction Act,* but from what I understand, adjudication is a statutory right under the Construction Act, and one that the parties cannot wriggle out of. The adjudication procedure under the Act has a very tight time limit. While this may help alleviate contractors' short-term cash-flow difficulties, the tight schedule makes it difficult for responding parties to review the many relevant documents in time, in order to effectively present their case. It also forces adjudicators to rush their decisions on what might be complex technical and legal considerations. All of which gives you some idea, I suppose, of why adjudication has been labelled 'rough justice'.

An advantage of statutory adjudication is that it can be effective in forcing the parties to find an early and sensible solution.

SUMMARY

It makes perfect sense for parties to find solutions to their differences/disputes, either between themselves through direct negotiation or with the help of a third party or third parties. The objective is (or should be) to avoid time-consuming and costly arbitration or litigation.

In the vast majority of projects I have worked on, disputes have been resolved when dispute resolution methods have been used, particularly, in my own experience, executive-level settlement negotiations, dispute boards or mediation.

If all attempts to find that sensible solution have failed, either in unsuccessful negotiations or after third-party intervention (the ADR process), the next and obvious step for the parties is arbitration.

---

* Housing Grants, Construction and Regeneration Act of 1996.

CHAPTER 8

When All Attempts to Settle Fail: Arbitration

Many disputing parties consider arbitration to be the most natural method of resolving the differences that they could not settle themselves.

A s a sitting arbitrator, I would welcome it if parties believed arbitration was the most natural method of resolving differences they were not able to settle by themselves. If this was indeed the case, I would be busy for the next twenty years – until I am ninety, at least, the age at which I might decide to retire.

But if I said that I believed arbitration was the *most natural* method of resolving disputes, I would be lying to myself. What I would really welcome – and I hope I have made this clear throughout my book – is for parties to avoid referring their differences, their disputes, to arbitration or to the courts. That said, if the parties have explored all other avenues, then, and only then, will I welcome parties engaging in arbitration for a just result – for the right decision.

Before discussing the arbitration process – a discussion I will try to keep reasonably short as readers are likely to be well versed in arbitration rules and procedures – I wish to introduce two interesting arbitration-related features I have encountered recently, both of which can contribute to settlement of disputes before or during the arbitration process. These are the 'shadow arbitrator' and 'third-party funding'.

THE SHADOW ARBITRATOR

'The core concept of a shadow arbitrator is to help a party and its counsel to better understand how an arbitral tribunal is likely to perceive and assess a single submission or the entire case.'*

If a tribunal is to be persuaded, it must understand the parties' positions – their arguments. If the issues are understood, the tribunal will be able to establish the facts and make the 'right' decision.

The shadow arbitrator is a term I discovered recently when I came across the article by Professor Jörg Risse in the Swiss Arbitration Association's bulletin that is quoted above. With Professor Risse's kind permission, I make reference to his views when examining the benefits of using a shadow arbitrator – the benefits that come from guiding clients and legal counsel in the drafting of submissions that will persuade a tribunal and that may even persuade the parties to settle.

I am very much in tune with Professor Risse when it comes to submissions being clear and concise, easy to understand and written in such a way as to hold the reader's attention. I have tried to convey this message throughout my book. Professor Risse discusses the important aspects of drafting submissions in what he calls his 'first level advice'. As he says in the article, 'If the arbitral tribunal does not understand a submission, even the best arguments are doomed to failure.' He continues: 'A similar serious problem arises when the writing itself is linguistically comprehensible but so complicated or boring that the arbitrator's mind loses interest in reading it.'

Professor Risse believes – and I agree – that the shadow arbitrator, being a practising arbitrator, can provide valuable feedback to, and guide, clients and legal counsel on how best to write submissions that will be readily understood. The shadow arbitrator can identify which passages of a submission need a special effort of concentration to be understood. Once this problem is identified, the passages can

---

* Professor Dr Jörg Risse: 'The shadow arbitrator: a mere luxury or real need?'. *Swiss Arbitration Association (ASA) Bulletin*, Volume 38 (2/2020, June).

be rectified to be more understandable. In Professor Risse's words: 'Knowing the problem is the first step towards solving it.'

Professor Risse's 'second level advice' relates primarily to the written word and how the shadow arbitrator can advise on improving the communicative quality of a submission or pleading. I like Professor Risse's psychology of prioritizing and structuring the issues (events): putting the strongest argument first, the third-strongest argument in the middle and the second-strongest argument at the end. In his words: 'First impressions count, last impressions stay.'

Professor Risse's 'third level advice' deals with the development of procedural strategy. Professor Risse asks:

> Why should a shadow arbitrator be able to assess the quality of an argument better than a competent counsel? The answer lies in the different perspectives. The shadow arbitrator's is that of the actual arbitrator. ... The counsel's close involvement in the creation of the arbitration file prevents him from having an unbiased view of the case and its merits.

It is for this very reason that I recommend clients engage independent consultants (experts) rather than using in-house personnel to develop their claims. As I explained in the preface, earlier:

> The reason I encourage contractors to engage an independent consultant/expert is for their objectivity. While contractors' in-house personnel might be perfectly capable of preparing claims, personalities, emotions and ownership of the issues all get in the way. Weaknesses may not be viewed objectively – or worse, they might be overlooked or ignored.
>
> In-house personnel often focus on what they perceive to be undisputed entitlement – their perceived rights – whether or not they have the evidence to support such perceptions. Weaknesses that are overlooked or ignored are weaknesses that will certainly be exploited by the other side. Unless they are recognized and addressed, such weaknesses will turn out to be the other side's strengths.

The shadow arbitrator is in a good position to predict the outcome of the case. He/she can substantiate his/her prediction just as objectively as the tribunal does in its final award. This applies equally to contractors' claims.

Professor Risse sums up the perfect shadow arbitrator:

> The shadow arbitrator shall reflect the thoughts and feelings of the arbitral tribunal; his job is to put himself in the arbitrator's shoes.
>
> The shadow arbitrator is not charged with seeking additional or better factual and legal arguments. But he must be able and willing to give candid feedback, even if this feedback does not please the party and its legal counsel.
>
> The shadow arbitrator should have significant experience as an arbitrator in order to understand how an arbitrator reads submissions and listens to a pleading and how an arbitral tribunal ultimately deliberates a case.

Professor Risse advocates, and again I agree with him, that in order to improve a party's chances of success in an arbitration, the shadow arbitrator should be involved in the arbitration process at an early stage – even before the party files its first submission. I wish to add that a shadow arbitrator should be engaged by clients at the time of preparing time and money claims, long before arbitration is on anyone's mind. The shadow arbitrator is the perfect person to guide the contractor and its claims consultant to prepare a claim that will persuade a tribunal, and in so doing persuade the employer that the claim has merit and should be taken seriously.

Professor Risse concludes his article by suggesting that

> Every arbitration is an investment in the outcome of the case, i.e., in the ultimate decision of the arbitral tribunal. The return on the investment is the awarded claim (claimant's perspective) or the dismissal of the claim (respondent's perspective). From this point of view, the involvement of a shadow arbitrator is also an investment. ... The investment in a shadow arbitrator, even if this investment is substantial, may well be one of the wisest.

Yet again, I agree wholeheartedly with the professor.

With the help of the shadow arbitrator, well-structured submissions and pleadings – written in a manner that will be readily understood by, and are likely to persuade, a tribunal – will and should alert the other side to the fact that such submissions and pleadings will more than likely succeed. Realizing the merits of such well-developed submissions will (should) persuade any wise opposition to enter into early settlement negotiations and try to find that sensible solution.

In my forty-plus years in the construction dispute management business, when developing strategies with clients and legal counsel, I have always tried to play the role of devil's advocate, examining all angles of every issue or dispute to ensure weaknesses as well as strengths are carefully considered before going into battle. My approach has often irritated my bosses and clients – sometimes they have asked whose side I was really on.

Quite recently, I had the pleasure – indeed the honour – of taking part in a mock arbitral hearing as the shadow arbitrator. It was an exhilarating and rewarding experience, and the feedback I received from legal counsel at the conclusion of the arbitration was extremely positive.

Having read Professor Risse's article, and looking back over my career, it would seem that all this time – while irritating my bosses and clients and without even realizing it – I was actually playing the part of the shadow arbitrator.

THIRD-PARTY FUNDING

What is third-party funding? What are the advantages of third-party funding?

In essence, third-party funding is a process in which an independent party (the funder) agrees to finance a party's arbitration (or litigation) costs in exchange for an agreed fee if the funded party wins its case.

While litigation funding has been in existence for quite some time, it is only in the past ten to fifteen years (as far as I am aware) that third-party funding has been used to finance construction arbitrations. The consensus among employers, contractors and legal counsel I have been in contact with is that third-party funding for construction arbitrations is here to stay and is growing in demand.

One of the attributes of third-party funding is that it gives a party access to justice when that party has problems with funding its case. In the limited arbitration cases I am aware of where one party, usually the claimant, has applied for and/or received third-party funding, this has been because of liquidity problems. That being said, funding out of necessity is not the only reason for third-party funding.

Funding out of choice has also become popular. Increasingly, financially sound parties prefer to use third-party funding for their arbitrations (or court cases) by choice, so that they can manage their cash flow and invest their millions elsewhere.

Prior to entering into a funding arrangement, the funder undertakes a full due diligence of the case. The funder examines the jurisdiction of the tribunal, does the merits and quantum analysis of the case, and researches enforcement perspectives. If the funder is satisfied with the prospects of the case and the parties are in agreement on the funding terms, they sign a funding agreement that sets forth *inter alia* the parties' rights and obligations, the budget allocated to the case and the funder's fee structure. The funder's only source of revenue derives from proceeds actually collected by its (successful) client. Typically, the funder's return is structured as a percentage of the claim proceeds.

A concern of some funders I have spoken to is safeguarding their fee if the parties negotiate a settlement before the conclusion of the arbitration. The involvement of funders in settlement negotiations can bring added value to the outcome, as the funders will be aware of the strength of the funded party's case as a result of the objective analysis undertaken in their due-diligence exercise. The funder's direct involvement in any settlement negotiations will be agreed to and expressly stated in the funding agreement.

Disclosure of the funding arrangement is something that has been debated for some time. There is no general obligation on a funded party to disclose the fact of being funded. However, given eventual conflicts of interests, the voices in favour of disclosure of funding arrangements and bigger transparency are growing. In my opinion, the funded party should disclose any third-party funding arrangement, and this should be disclosed early in the arbitration process. Early disclosure can have distinct benefits. The fact that a funder believes it is worth investing in a party's case will give that party confidence. More importantly, the other party should recognize that their opponent is likely to have a case that may well persuade the tribunal. This should be an incentive to settle.

THE ARBITRATION PROCESS

Choosing the lawyers to manage the arbitration

The choice of legal counsel to represent the parties in arbitration, and for that matter in litigation, is of vital importance and must be considered carefully by parties contemplating referring their dispute to arbitration. I believe it goes without saying that it is essential for legal counsel to be experienced in construction and infrastructure projects as well as in the law governing the contract (i.e. common law or civil law). Legal counsel should also have experience in managing construction arbitrations for its clients.

I have recently experienced arbitrations in which parties have been represented by their in-house legal teams. Naturally, this can be a financial advantage for the party, but I question whether in-house lawyers can be as objective in their strategy and pleadings as external counsels.

In either case, whether one uses external or in-house counsel, I recommend parties seriously consider engaging a shadow arbitrator to assist in developing their case.

The role of the expert in the settlement of disputes

The role of an expert, from my personal experience, falls into two distinct categories.

(i) *The independent expert:* providing objective advice to clients on the merits of their case and recommending how best to proceed to resolution.

(ii) *The expert witness:* engaged when the parties are already committed to arbitration, providing an independent and impartial opinion on aspects of the case, usually of a technical and/or complex nature, to assist the tribunal's understanding of particular issues.

The independent expert

I have included the role of the independent expert within the arbitration process because the opinion and recommendation of the expert may encourage their client to seek a solution rather than continue with the arbitration.

When the expert is asked to provide an opinion and recommendation at an early stage, before disputes manifest themselves and before the idea of arbitration is born, the chances are that the parties will settle and relationships will be preserved. In many ways, the role of the independent expert mirrors that of the shadow arbitrator.

The expert witness

The expert witness's role is to help the tribunal understand technical complexities that are within the expert's field of expertise: technical matters that may be outside the knowledge or expertise of members of a tribunal.

It is rare, in my experience, for expert witnesses to contribute to settlement of disputes during the arbitration proceedings or before the tribunal's final award. Having said that, in situations where the 'opposing' experts have met and agreed on areas of common ground, the parties may be encouraged to attempt settlement on any remaining differences. It is also possible for parties to settle after the hearing, when the expert witness testimonies have convinced one or other party on the merits of the other party's case.

EFFICIENT MANAGEMENT OF THE ARBITRATION

There was a time when arbitration was favoured over litigation because it was seen to be quicker and cheaper. This has not always been the case in my experience. When I was first introduced to the world of international arbitration in the mid 1990s, arbitration was a time-consuming and costly process. This appears to have been recognized by most arbitral institutions and, as such, parties, legal counsels and tribunals are being encouraged to expedite the process to achieve efficient and cost-effective arbitrations. This seems to be working.

Once appointed, the tribunal plays a key role in the efficient management of the arbitration. As noted in the Rules on the Efficient Conduct of Proceedings in International Arbitration (the Prague Rules*):

> It has become almost commonplace these days for users of arbitration to be dissatisfied with the time and cost involved in arbitration proceedings.
>
> One of the ways to increase the efficiency of arbitral proceedings is to encourage tribunals to take a more active role in managing the proceedings.

I totally agree with the Prague Rules Working Group. If managed well by the tribunal and with the cooperation of the parties, arbitrations will continue to become more efficient.

The tribunal can also play a key role in encouraging the parties to settle their disputes at any time during the arbitral process. This encouragement to settle should start during the first meeting between the parties and the newly formed tribunal.

The arbitration agreement

Efficient and effective management of arbitrations begins with the arbitration agreement.

When the parties to a construction project wish to have their disputes finally settled by arbitration, their contract must include a

---

*Rules on the Efficient Conduct of Proceedings in International Arbitration (Prague Rules), published in 2018. According to the preamble on page 3, the Prague Rules are intended to provide a framework and/or guidance for arbitral tribunals and parties on how to increase efficiency of arbitration by encouraging a more active role for arbitral tribunals in managing proceedings.

clause (the arbitration agreement) stipulating that all disputes arising out of or in connection with the contract *shall* be finally settled
by arbitration. The arbitration agreement must be clear and free of
ambiguity. Vague wording can compromise the agreed dispute resolution process and may result in the disputes being referred to the
courts, rather than to arbitration.

The arbitration agreement, if it is to be effective, must include the
following:

(i) a stipulation that all disputes arising out of or in connection
with the contract *shall* be finally settled by arbitration;
(ii) the arbitration rules under which the arbitration is administered;
(iii) the number of arbitrators forming the tribunal – either three
arbitrators or a sole arbitrator;
(iv) the 'place' or 'seat' of the arbitration; and
(v) the language of the arbitration.

If an arbitration agreement is properly drafted – and this is not
rocket science, as seen above – the parties will be bound by this
agreement to have their disputes referred to arbitration. If either
party ignores the arbitration agreement and submits the dispute to
the courts for resolution, the courts are likely to reject the case on the
grounds that the parties have expressly agreed to have their disputes
resolved by way of arbitration.

The place or seat of arbitration

The place or seat (I will use 'seat' from now on) of the arbitration stated
in the arbitration agreement dictates the law that will govern the procedure of the arbitration. It is often the case that the law of the contract
the parties have entered into will not be the same as the law governing
the arbitration. For example, a contract for works being caried out in,
say, Qatar will be governed by Qatari law. If the seat of the arbitration
is, say, Paris, then the law governing the arbitration will be French law.

The seat of the arbitration – Paris in the example just given – does
not necessarily mean that meetings and hearings have to be held in
Paris. Meetings and hearings can be held in any country the parties
and tribunal agree to.

> In a recent arbitration I was involved in, the project was in Doha and the seat of arbitration was Zurich. Meetings between the parties and tribunal took place in Zurich, Paris and London. The final hearing was held in Geneva.

The seat and the respective governing laws can be confusing for parties unaccustomed to some of the finer points of arbitration. It took me a while to understand the implications of the seat, place and applicable laws.

Request for arbitration: answer to the request

To begin the arbitration process, one of the parties, the Claimant, will submit a Request for (Notice of) Arbitration to the Secretariat of the Institution named in the arbitration agreement. The Respondent will submit its Answer (Reply) within the time specified in the Institution Rules. Without exception, all arbitration rules stipulate that the Request and Answer shall provide, in addition to other details:

(i) a description of the disputes (claims) being referred to arbitration; and

(ii) an indication of the monitory value of the claims and counterclaims.

Appointment of the arbitrators

The appointment of the tribunal should be a straightforward procedure. In cases where the tribunal is made up of three arbitrators, under most arbitration rules, each party will nominate a co-arbitrator. Once confirmed, the co-arbitrators will then agree on the third arbitrator, who will be chairperson (or president) of the tribunal. If the co-arbitrators cannot agree on the third arbitrator, he or she will be appointed by the arbitration court.* In the case of a sole arbitrator, the parties will attempt to agree on that arbitrator. If the

---

*The 'court' being the institution governing the arbitration. For example, the International Court of Arbitration of the International Chamber of Commerce.

parties are unable to agree on the sole arbitrator, this arbitrator will be appointed by the arbitration court.

Composition of the tribunal
In the early days of my involvement in international construction arbitrations, tribunals were almost exclusively made up of lawyers. This observation was reinforced by my friend and mentor Christopher Seppälä, as told in the story below.

After being involved with a few international arbitrations, I decided to look into studying to become an arbitrator. When visiting Paris around that time, I was enjoying lunch with Christopher, who had taught me everything I knew about arbitration and how to draft submissions that will persuade a tribunal.

Over lunch, I told Christopher of my plan to study to be an arbitrator. I will never forget his response. 'When we are ready to nominate a co-arbitrator for our arbitrations, in the great majority of cases we, and in my experience most other law firms, will nominate a lawyer as an arbitrator. A technical person, like yourself being a quantity surveyor, or an engineer, would only be nominated in exceptional cases, where technical issues appear predominant.'

Christopher went on to explain: 'This is because a non-lawyer arbitrator is likely to be less familiar with arbitration law and less able to deal with legal issues that may arise, which legal issues could decide the case. That being said, there are some non-lawyer arbitrators who have acquired experience in arbitration and who, because of their skill and experience, can be as effective as lawyer arbitrators. But they remain the exception rather than the rule.'

Christopher's observations were a bit of a setback for my arbitration plans, but it turned out well for me: I became a mediator instead. As I have mentioned earlier in this book, being a mediator has proved invaluable when helping parties – my clients – find that sensible solution.

In more recent times I have witnessed more balanced tribunals, often made up of two technical arbitrators (usually an engineer and a quantity surveyor) along with one lawyer. This shift to a more balanced tribunal has been particularly fortunate for me, as discussed below.

> Having taken on board my friend Christopher Seppälä's observations and shelved my plans to become an arbitrator, it came as a total surprise when I received a call from another lawyer who I had worked closely with for a few years. He calmly told me that he had nominated me as co-arbitrator for a case on which he was representing the Respondent.
>
> 'I'm flattered,' was my immediate response, 'but you know I'm not an arbitrator and that I've had no training towards becoming an arbitrator.' He replied: 'I know all that Wayne, but I've worked closely with you for quite some time. You think like an arbitrator. I believe you will be a good arbitrator.'

The tribunal in my first case as arbitrator included two lawyers plus one quantity surveyor: me. Since that initial experience as an arbitrator, I have been appointed several more times, and in every case the tribunal consisted of two technical arbitrators and one lawyer, usually the president (chairperson). This balanced structure has worked extremely well, and I believe it is the right balance for construction arbitrations.

Since becoming an arbitrator, I have had the good fortune of working alongside excellent arbitrators and have learnt a considerable amount from each of them, including efficient management of the arbitration and rendering the awards within the time frames stipulated in the various arbitration rules.

Case management conference

The case management conference (meeting) is more often than not the first time the parties and their respective counsels meet with members of the tribunal.

In this meeting, which is important for setting the tone for an efficient arbitration, the following matters will be discussed and agreed to by the parties and the tribunal:

   (i) terms of reference,
  (ii) procedural timetable,
 (iii) the pleadings,
 (iv) expert witnesses,
  (v) witnesses of fact and
 (vi) disclosure of key documents.

Terms of reference

Soon after the arbitration file has been forwarded to the tribunal by the institution's secretariat, the tribunal will draft the terms of reference.* The arbitration file will usually consist of, among other documents, the Request for (Notice of) Arbitration (by the Claimant) and the Answer (Reply) to the Request (by the Respondent). The 'Request' and the 'Answer' will summarize the parties' respective claims and relief sought – the relief being, for example, extension of time and the amounts (monetary value) of the claims and counterclaims.

The draft terms of reference will state the issues to be decided upon in the arbitration, as gleaned from the arbitration file. The initial draft will be sent to the parties for agreement/comment before the case management conference. During the case management conference, the terms of reference will be discussed, amended as necessary, agreed to and signed by the parties and members of the tribunal. Once the terms of reference are signed, no additional matters or claims will be considered in the arbitration unless agreed to by both parties and the tribunal.

Procedural timetable

Before a party makes the decision to refer their dispute to arbitration, both parties will have been through the process – at times a lengthy process – of the preparation, submission and rejection of claims, failed negotiations, executive attempts to settle and third-party

---

*I have observed that not all arbitral institutions state the requirement for terms of reference. In a recent case for which I was part of the tribunal there were no terms of reference, and the only document agreed to during the 'case management meeting' was the procedural timetable.

(ADR) resolution procedures. All details of the claims and coun-terclaims will have been closely examined and all supporting docu-ments will have been scrutinized. The issues in dispute will be well known and understood by both parties.

Because, as mentioned above, the issues in dispute will be well known and understood by the parties before entering arbitration, I see no reason for either party to request more time to submit their written pleadings than is stipulated in the various rules.

The focus of the tribunal – and, in particular, the parties and legal counsel – should be on completing the arbitration as quickly and efficiently as possible.

While the emphasis is on efficiency, the procedural timetable proposed by the tribunal must be realistic and achievable. The pro-posed timetable will be carefully considered by and agreed to by both parties during the case management conference.

In summary, the steps to be taken and the matters to be dealt with in the arbitration and set out in the procedural timetable will, in most cases, include the following.

(i) Dates for submission of the parties' pleadings: Claimant's state-ment of claim, Respondent's statement of defence and coun-terclaim, Claimant's reply/answer to the counterclaim, and Respondent's rejoinder (response to Claimant's reply/answer to the counterclaim).

(ii) Final date for disclosure/exchange of specific documents requested by the parties and agreed to by the tribunal.

(iii) If expert witnesses are to be engaged, dates for submission of the experts' reports, dates for meetings between the experts, and final date for any joint report/statement of the experts.

(iv) Dates for submission of signed statements from witnesses of fact.

(v) Date for submission of hearing bundles.*

(vi) Date(s) and location of the hearing and agreement as to whether the hearing will be live or virtual.

---

* Hearing bundles are the files containing all documents relating to the arbitration – to give easy access to all parties during the hearing.

The tribunal may revise the timetable from time to time to ensure the parties have sufficient opportunity to properly plead their respective cases.

The pleadings (submissions)

For an efficient arbitration, the parties' pleadings (submissions) should be short. The principle for preparing contractors' claims that we discussed in chapter 5 applies equally to the parties' written submissions in arbitration. To persuade a tribunal, the parties' submissions must be clear, concise and credible. This principle also applies to the exhibits – the supporting documents: the evidence.

A colleague of mine was involved in an arbitration recently in which both parties submitted millions of pages as exhibits: so-called evidence. The tribunal was not impressed. To ensure justice was done and the correct decisions were made, the tribunal had to examine every document. As a result it took the tribunal almost two years to deliberate over and finally render the award.

To avoid excessive and mostly unnecessary paperwork, the tribunal should be firm with the parties during the case management conference regarding excessive documentation. This I believe is an important and essential part of the tribunal's efficient management of the arbitration.

Document disclosure

Unlike litigation in the United Kingdom and, as I understand it, in the United States, where the disclosure process can be cumbersome and time-consuming, disclosure in arbitrations should in my opinion be limited to only those documents directly related to the issues. From my experience as an arbitrator, parties have usually provided all supporting documents with their submissions. Seldom have I experienced parties demanding further documents be disclosed by the other side. This is how it should be, I believe, and it is in line with the growing demand for more efficient arbitrations.

Expert witnesses

I have briefly discussed the role of the expert witness earlier in this chapter so I will not repeat my observations here except to say that expert witnesses can be crucial in complex construction arbitrations, particularly if the tribunal is comprised only of lawyers, who may have little or no technical expertise.

The tribunal should, during the case management conference, establish (i) if the parties intend appointing expert witnesses, (ii) the issues to be opined on by the experts, and (iii) what the experts can contribute to the arbitration.

If the tribunal is made up of a balance between technical and legal arbitrators, it may not be necessary to have experts explain the technicalities to arbitrators who are fully versed in those technicalities. In such a situation, it may be wise for the tribunal to advise the parties against involving experts. This will save both time and money.

On the other hand, if it is decided that experts will add value to the proceedings, the tribunal will need to be clear on the issues to be opined on and the timing of the expert reports, meetings and joint reports.

Witnesses of fact

Witnesses of fact, also known as fact witnesses, are in most instances employees of the parties. These witnesses can play an important role in persuading a tribunal – particularly when they are telling the truth. When witnesses are not being truthful, or are simply toeing the party line, the tribunal will recognize this during the hearing. Witnesses who are not being totally honest might well persuade the tribunal, but not in the way that they or their employers would like.

TRIBUNAL TO ENCOURAGE SETTLEMENT

During the case management conference and at any time between the conference and the hearing, the tribunal should encourage parties to continue talking to each other whenever the opportunity arises. The parties should be trying to find a way to reach a sensible settlement.

On a case for which I was a member of the tribunal, it was clear that the matter should never have been referred to arbitration. Neither the issues nor the amounts claimed and counterclaimed were sufficient to warrant arbitration. The tribunal asked the parties and their legal counsels to attend a preliminary meeting. We expressed our concerns and encouraged the parties to talk each other – to find a sensible solution. The parties were asked to report back to the tribunal within a specified time.

Within a few days the parties reported that they had reached agreement and provided the tribunal with the details. All that remained for the tribunal was to render a consent award.

THE HEARING

While the parties should take an active role in the arbitration process, as is their right, when it comes to the hearing, the tribunal is in charge of proceedings. The tribunal will, naturally, need the parties' full cooperation to ensure the hearing is managed in the most efficient and cost-effective way.

In good time before the hearing, the tribunal and the parties should agree on the duration of the hearing, ensuring the parties are comfortable with being able to present their respective cases and to have their witnesses (expert and fact witnesses) heard. The tribunal should always be conscious of the time and cost implications of a prolonged hearing and should manage the hearing accordingly.

Physical (oral) hearings

Until the arrival of the Covid-19 pandemic in early 2020, oral hearings were held in physical locations with members of each party, the witnesses and, of course, the tribunal attending in person. Since the arrival of the pandemic, many hearings (as well as case management conferences and other meetings) have been held remotely, or virtually, by way of Zoom or other forms of video conferencing.

Irrespective of whether the hearing is held physically (in person) or virtually, the agenda usually follows the same pattern, starting with legal counsels' opening statements, then witnesses giving evidence and finally closing statements by legal counsel.

Two important features of arbitral hearings are the opening and closing statements, which I feel need to be briefly discussed.

Opening statements

Opening statements are important as they crystallize each party's case for the tribunal. They are the 'executive summary' of the pleadings (submissions). The statements should be clear and concise, summing up the key aspects of the party's case, making reference to the evidence as appropriate.

The tribunal will allocate equal time to each party for presenting their case at the hearing. This time allocation will include the opening statements, the time for witnesses (fact and expert witnesses) to be heard and cross-examined, and closing statements. The more concise the opening statement, the more time each party will have for their witnesses to be heard and for the closing statements.

> In one arbitration I was party to, the hearing was to be conducted in one 'eight-hour day', each party being allocated four hours to present their case. Legal counsel for the Claimant took up most of his client's four-hour allocation in a drawn-out and tedious opening statement. He insisted on 'reading' his client's statement of claim almost word for word, ignoring the presiding arbitrator's repeated reminders of the need to keep the opening statement short.
>
> As a result, the tribunal was forced to reorganize the hearing and time allocations. Instead of the hearing being concluded at 6 p.m. as planned, the tribunal finally closed the hearing at 10 p.m., with closing statements being submitted as post-hearing briefs a week later.

The Claimant was not pleased with the manner in which counsel presented their case and was even less pleased when the tribunal awarded costs relating to the extended hearing and post-hearing briefs in favour of the Respondent. In his opening statement, legal counsel missed a golden opportunity to highlight the strengths of his client's case. As noted by Professor Risse (in the 'shadow arbitration' section earlier), first impressions count. The first impression certainly counted with the tribunal, but not in the way the Claimant wished.

Closing statements
Unless the case is complex and the hearing has taken up the fully allocated time, legal counsel should be able to sum up their client's case in an oral statement at the close of the hearing. I have noticed that in some cases the tribunal has instructed legal counsels to submit written post-hearing statements (briefs). Invariably, in my experience, the post-hearing briefs do no more than summarize what has been written in the pleadings and stated during the hearing. I believe that this can just as easily be conveyed in short closing statements.

In a case I was involved with (as a party) the tribunal instructed the parties' legal counsels to exchange post-hearing briefs and then reply to these briefs. The time given for the parties to exchange these two sets of documents was six months after the end of the hearing. It took the tribunal another year to render its award. It was not a complex case.

In this particular case, the tribunal prolonged the arbitration time, which in turn contributed to a considerable increase in the arbitration costs. All of this was unnecessary in my opinion.

Final encouragement to settle before the award
As a final statement before closing the hearing, the tribunal may – and in my opinion should – remind and encourage the parties to communicate in an effort to settle their disputes. At this stage in the arbitration, both sides should have a fair understanding of the other side's strengths and their own potential weaknesses.

Documents only

An alternative to a physical, or even a virtual, hearing is for the tribunal to render an award on documents only. Most arbitration rules

I have worked with provide for documents-only awards, without the need for a hearing.

Documents-only arbitrations do have the advantage of being more cost-effective. They are also encouraged under the Prague Rules, article 8.1 of which states:

> In order to promote cost-efficiency and to the extent appropriate for a particular case, the arbitral tribunal and the parties should seek to resolve the dispute on a documents-only basis.

It will not surprise me if documents-only arbitrations become the norm as an alternative to 'virtual' arbitrations in the aftermath of the Covid-19 pandemic.

THE AWARD

In the interest of the parties and in line with the growing desire for efficient and cost-effective arbitrations, the tribunal must endeavour to render its award as soon as possible after the hearing or, in the case of documents-only arbitrations, after the final documents have been exchanged by the parties. The rules of most institutions require the award to be rendered within six months of the end of the hearing or, in the case of the International Chamber of Commerce (ICC) Arbitration Rules, within six months of the signing of the Terms of Reference. Six months should be more than enough time for any tribunal to render an award. Under the Qatari Civil and Commercial Arbitration Law (Law No. 2 of 2017), unless otherwise agreed by the parties, the award is to be rendered within one month (or a maximum of two months) from the end of the hearing.* This more limited time is also achievable.

A large part of the award can be written before the hearing. Even the parties' positions and arguments relating to the issues, which will be known to the tribunal in advance of the hearing, can be written in 'draft'. The only sections of the award that will not be

---

*Article 31.5 of the Qatari Civil and Commercial Arbitration Law (English translation).

drafted before the hearing will be the tribunal's findings. All that is needed to finalize the award after the hearing and, if applicable, post-hearing briefs will be some fine-tuning of the parties' positions and arguments, and the drafting of the tribunal's findings, decisions and reasoning.

I was recently astounded by, and delighted with, a judge presiding over a Technology and Construction Court UK* case in London and with a sole arbitrator in a case in Qatar: they both rendered their decision/award in less than a week after the hearings.

In the various arbitrations in which I was co-arbitrator in Qatar, I am pleased to reveal, in all modesty, that the awards were rendered well within the time frame stipulated under the Qatar Civil and Commercial Arbitration Law.

SUMMARY

The priority for the tribunal, and for the parties, is to ensure an efficient and cost-effective arbitration. A sound starting point for an efficient arbitration is the arbitration agreement.

If the parties intend their disputes to be settled by way of arbitration, this must be made clear in the arbitration agreement. If the agreement is not absolutely clear that disputes shall be settled in arbitration, this may leave the door open for a party to refer the case to the courts – if they feel it will be to their advantage to have their case heard in the local courts.

A balanced tribunal comprising at least one technical member, and preferably two, plus a lawyer (as the presiding arbitrator) is, in my opinion, ideal for construction-related arbitrations. Having one or two technical arbitrators on the tribunal can eliminate the need for technical expert witnesses and, as such, saves time and is more cost-effective.

For an efficient and cost-effective arbitration, the tribunal has the important task of managing the process. A well-managed arbitration

---

*The Technology and Construction Court UK deals primarily with technology and construction disputes. It was previously known as the Official Referees Business, Queen's Bench Division.

– from the initial case management conference through guiding parties to submit concise and well supported pleadings, proposing a realistic but efficient procedural timetable, and managing the hearing effectively – will ensure a smooth and efficient arbitration.

From the parties' perspective, the most important part of their arbitration is the result: the tribunal's award. The tribunal will scrutinize every submission and every exhibit presented by the parties with great care. The tribunal will study expert and fact witness statements and listen carefully to legal counsel presentations and witness testimonies during the hearing. This careful attention to detail will ensure that the tribunal's award is correct and just.

Throughout the process, the tribunal should, whenever appropriate, encourage the parties to engage in constructive dialogue and try to find a sensible solution. The parties should be reminded that they can settle at any time during the arbitration process. Most arbitrators will be delighted to be asked to render a consent award.

Parties can also be encouraged to settle their disputes if either party, or indeed both parties, engage a shadow arbitrator at an early stage. The shadow arbitrator, being a practising arbitrator, is in a position to advise a party and its legal counsel on whether or not a tribunal will be persuaded. If the shadow arbitrator feels the party has a potentially weak case, this may encourage that party to return to the negotiating table and try to settle.

I wish to close this chapter by referring back to third-party funding. If a party to an arbitration is in need of, or decides to arrange, funding for whatever reason, I recommend that the party discloses to the other party and to the tribunal that it is being funded by a third party. Early disclosure of third-party funding has two advantages: (i) the funder, having carried out its due diligence, will only agree to fund the case if it believes the case has merit and is likely to win; and (ii) by disclosing the use of a third-party funder, the other party should recognize that their opponent's case may be good enough to persuade the tribunal, which should be an incentive to settle.

CHAPTER 9

Conclusion: Seeking Sensible Solutions

Disputes are inevitable in international construction. If our projects are to be completed successfully and relationships maintained, all parties must work together to find sensible solutions to disagreements and potential disputes.

The theme I have tried to convey throughout this book is that all parties involved in construction projects should be looking for sensible solutions to their problems. There will always be differences of opinion and arguments over progress and money – differences that often lead to disputes, to conflict, which unless resolved sensibly can end in arbitration or in the courts.

In my book I deal with dispute management and avoidance in the following general stages:

(i) managing projects and people, to avoid disputes;
(ii) preparing clients' claims to persuade a tribunal;
(iii) negotiating to find sensible solutions; and
(iv) managing disputes when all attempts to settle fail.

MANAGING PROJECTS TO AVOID DISPUTES

In the opening chapters of my book I focused on wise management of projects and contracts and, most important of all, managing people to achieve common goals: executing and completing projects to a high standard, on time and within budget. Achievement of these basic goals can prevent, or at least minimize, disputes.

People can be a risk when dealing with day-to-day issues that, if not well managed, can incite disputes. Senior management – executives of both the employer and contractor – should constantly monitor the atmosphere 'on the ground', and when necessary they should take early action to prevent differences escalating to disputes – to conflict.

Communication is the key to preventing or minimizing disputes. All people involved in construction projects, at whatever level of management, should communicate on a regular basis. When I say communicate, I mean talk to each other, listen and try to understand each other's positions. As I have mentioned a few times in this book, understanding the other party's position does not necessarily mean agreeing with it, but it can be instrumental in finding that sensible solution.

Understand your contract. This cannot be overemphasized. When all terms and conditions, all provisions – in particular rights, obligations and procedures – are understood and complied with by all parties, unnecessary disputes will be prevented, and a successful project can be achieved.

PREPARING CLAIMS TO PERSUADE A TRIBUNAL

The first essential step in developing a persuasive claim is the investigation. From the findings and opinions formed during our investigations, a strategy for a credible claim can be developed. A well-prepared claim should encourage the employer to make a fair assessment and to reach an early and sensible agreement.

From the detailed research undertaken during the investigation process, we will have established most of the facts and located the evidence to support each element of the claim. This will place us in a good position to start writing a claim that will persuade a tribunal.

As emphasized in chapter 5, for the claim to be persuasive, it must fulfill the following criteria.

- It must be clear and easy to understand, particularly by the employer's executives and later, if necessary, by a tribunal.
- It should be comprehensive but concise. The report must hold the reader's attention at all times. When the report is unnecessarily

protracted, the reader is easily distracted, or becomes bored. When this happens, our message is lost.

- It must be kept simple. Do not overwhelm the readers with complex descriptions or technical jargon. It will be difficult, if not impossible, to hold their attention. They will quickly lose interest.
- It should be objective. Be respectful. The tone of the claim report is important: objective arguments and respectful phraseology can be persuasive.
- It must be credible. All statements and arguments must be supported by the evidence.
- It should be realistic. The claims must be based on actual and not perceived facts. Cost claims must not be exaggerated: if they are, this will jeopardize credibility and be an obstacle to settlement.

The appendices to the claim report – the engine room – is where the details – the analyses, calculations and documentary evidence – are located. These are the documents that will support the contractor's claim.

NEGOTIATING TO FIND THAT SENSIBLE SOLUTION

Preparation before negotiations is essential. Having worked with our client through the investigation stage and then during the development of the claim, the strengths and potential weaknesses of our client's case will be known. Our client must have realistic expectations and be prepared to accept an outcome that they can live with.

Before entering into negotiations, a realistic 'high–low' settlement range should be agreed to with our client: a range between the best-case scenario and the lowest deal our client is prepared to accept. If the other side is willing to settle within my client's realistic expectation range, agreement will be reached.

MANAGING DISPUTES WHEN ALL ATTEMPTS TO SETTLE HAVE FAILED

In the projects I have been involved with in over the years, the vast majority of differences and disputes are resolved directly between the parties without having to resort to alternative dispute resolution such as mediation. It is rare for disputes on projects I have worked on to be referred to arbitration or the courts.

Mediation is becoming popular in construction and has recently become mandatory in many forms of construction contracts as a prerequisite for referring disputes to arbitration. When both parties are willing to settle, mediation will invariably help them find that sensible solution.

The inclusion of dispute adjudication boards (DABs) in construction contracts, as contained in the FIDIC forms of contract, has proved to be successful in encouraging parties to resolve their differences, either before or soon after receiving a DAB's decision. This works best with 'standing' DABs, where the DAB members meet with the site management on a regular basis. During these site meetings, DAB members can use their influence to reconcile the parties' differences on an ongoing basis.

When all means of settling disputes have failed, we are left with arbitration or, in some cases, the courts. The arbitration process has been dealt with quite extensively in the previous chapter and I will not repeat myself except to recommend that arbitration is to be avoided where possible.

AVOIDING ARBITRATION

My theme throughout this book has been for parties to construction projects to find sensible solutions to their differences – to their disputes. They should avoid resorting to arbitration for a final resolution. Apart from the damage that can be done to relationships by doing so, parties should consider the cost of arbitration – costs such as institution charges, arbitrators' fees, legal counsel fees, expert witness fees and travelling and accommodation costs when meetings and hearings are conducted in various parts of the world.

A cost that is difficult to quantify is management and staff downtime. The personnel involved on the project that is subject to arbitration lose an enormous amount of time, away from the projects they are working on, because of their ongoing involvement in the arbitration. This hidden cost can be considerable.

Another – and, in my opinion, equally important – factor in avoiding arbitration is the damage it can do to relationships. Maintaining good relationships with all parties is vitally important. It is essential for overcoming difficulties and ensuring the successful execution and completion of a project.

AND FINALLY, THE SHADOW ARBITRATOR

As a sitting arbitrator, I firmly believe that engaging a shadow arbitrator will contribute significantly to assisting parties find that sensible solution, thereby avoiding unnecessary disputes and costly arbitration.

With the help of a shadow arbitrator, particularly if one is engaged at an early stage, claims and arbitration pleadings/submissions can be written in a way that will persuade a tribunal. Submissions that are good enough to persuade a tribunal will, and should, alert the other side, the employer, to the fact that such submissions are likely to succeed. Realizing the merits of well-developed claims and pleadings, a wise employer will be persuaded to enter into early settlement negotiations with the contractor to find that sensible solution.

Further Reading

Bailey, Julian. 2020. *Construction Law*, 3rd edition. London Publishing Partnership, London.

Baker, Ellis, Ben Mellors, Scott Chambers and Anthony Lavers . 1999. *FIDIC Contracts: Law and Practice*, 3rd edition. Sweet & Maxwell, London

Blattner, Curt. 2020. *The Heartbeat of Excellence*. LID Publishing, London.

Centre for Effective Dispute Resolution. 2016. *Model Mediation Procedure*. CEDR, London.

FIDIC. 2020 (April). COVID-19 Guidance Memorandum to Users of FIDIC Standard Forms of Works Contract. FIDIC.

Gaitskell, Robert (ed.). 2017. *Keating Construction Dispute Resolution Handbook*, 3rd edition. ICE Publishing, London.

Goldsmith, Lucy, and Andrew Stephenson. 2021 (June). Time for Australia to embrace dispute resolution boards? Corrs Chambers Westgarth, Sydney, Australia.

International Chamber of Commerce. 2018 (March). *Commission Report: Controlling Time and Costs in Arbitration*. ICC, Paris.

International Chamber of Commerce. 2019. *Commission Report: Construction Industry Arbitrations, Recommended Tools and Techniques for Effective Management*. ICC, Paris.

Latham, Simon. 2021. *The Third-Party Litigation Funding Law Review*, 4th edition. Law Business Research, London.

Prague Rules. 2018. *Rules on the Efficient Conduct of Proceedings in International Arbitration*.

Risse, Jörg. 2020. The shadow arbitrator: mere luxury or real need? *ASA Bulletin*, Volume 38(2).

Royal Institution of Chartered Surveyors. 2014. *Mediation – RICS Guidance Note*, 1st edition. RICS, Coventry, UK.

Royal Institution of Chartered Surveyors. 2014. *Surveyors Acting as Expert Witnesses: RICS Practice Statement and Guidance Note*, 4th edition. RICS, Coventry, UK.

Society of Construction Law. 2017 (February). *Society of Construction Law Delay and Disruption Protocol*, 2nd edition. SCL, Hinckley, UK.

About the Author

Wayne is a chartered quantity surveyor, arbitrator and mediator with more than fifty years of international experience, having worked in the capacity of consultant to clients, contractors and subcontractors on major building and infrastructure projects in Africa, the United Kingdom, Europe, Asia and the Middle East.

For the past forty years Wayne has focused on contract, dispute and risk management, and primarily on dispute avoidance. Wayne's main role in dispute management is case-managing arbitration and court cases, and he has been appointed as an arbitrator, a mediator and an expert witness on various international construction disputes. In undertaking contract management tasks, Wayne's focus has been to guide clients towards sensible and clear communication, nurturing relationships and seeking early solutions to problems, with the primary goal being to avoid unnecessary conflict.

Wayne is now based in Lucerne, Switzerland, where he provides dispute management support to the international construction industry, including arbitration, mediation, claim evaluation and/or preparation, dispute avoidance and amicable settlement strategies. Wayne also advises third-party funders on the merits of potential client arbitration cases (due diligence). As a result of his appointments as arbitrator, Wayne has also acted as a 'shadow arbitrator', providing feedback to legal counsel and clients on those aspects of their submissions that may or may not persuade a tribunal.

Wayne has lectured at the Universities of Hong Kong and KwaZulu-Natal in South Africa; presented conference/seminar papers in Greece, Thailand and Qatar; run in-house workshops and training programmes for organizations based in the United Kingdom, Thailand, Bulgaria and Qatar; and has had articles published in construction journals in Qatar and the United Arab Emirates. Topics covered by Wayne include dispute management, arbitration, mediation, dispute boards, risk and construction management.